VHF RA1
Ten Golden

1 Listen and think before transmitting.

2 When two ships are establishing contact, it is normally the ship called which names the channel to be used for working. Always await acknowledgement before changing channel.

3 Although Channel 16 is the distress and safety channel, it is also used for calling. However, always call on a working channel if one is known to be watched.

4 Except for distress and urgency traffic Channel 16 must not be used for the exchange of messages.

5 Use low power for transmitting, whenever it will give satisfactory communication.

6 When a call is complete, and subsequently during an exchange of messages, a station invites a reply by saying: 'Over'.

7 Where a message is received and acknowledgement only is needed, say 'Received'. Where it is necessary to confirm that the information is correct and/or understood, say: 'Received – Understood' and repeat the substance of the message if considered necessary.

8 The end of work is indicated by each station adding 'Out' at the end of its last reply.

9 When conditions are good, abbreviated procedures may be used.

   (i) Repetition of words and phrases should be avoided.
   (ii) At the end of work only one station need use 'Out'; the other station need not reply.

10 When the traffic is complete ensure that the press-to-speak switch is properly released and that a carrier wave is not being transmitted when the microphone is in its cradle.

YACHT SIGNALLING
Bernard Hayman

NAUTICAL BOOKS
MACMILLAN LONDON

Front cover photographs:
Top left: *East Anglian Daily Times*
Top right: Patrick Roach
Bottom left: Bernard Hayman
Bottom right: Patrick Roach

Copyright © Bernard Hayman 1983

All rights reserved. No part of this publication may be reproduced or transmitted in any form, or by any means, without permission.

ISBN 0 333 32528 1

First published in Great Britain 1983 by
NAUTICAL BOOKS
an imprint of Macmillan London Limited
4 Little Essex Street, London WC2R 3LF

Associated companies throughout the world

Printed in Hong Kong

Contents

Author's preface   v

Part 1 Radio telephony
   Introduction   3
1. VHF R/T is really a giant party line   7
2. Rules and regulations   22
3. Standard Maritime Navigational Vocabulary   33
4. Summary of recommended procedure and techniques   41
5. Calling procedures   45
6. Distress and urgency procedure   58
7. Jargon   69
8. HM Coastguard and communications   73

Part 2 – Code or plain language.   Flags, light and sound
   Introduction   81
9. What type of signal   82
10. Methods of signalling   91
11. Signalling procedures   103
12. Ensigns and special flags   113
13. Signalling and yacht racing   124
14. Navigation lights and shapes   132

Part 3 – Emergency and other special topics
15 Distress and urgency   153
16 Radar reflectors, life-saving, port operation
    and night vision   171

Appendices
A Useful abbreviations   180
B VHF R/T anomalies   182
C Citizens' Band R/T   184
D Medium frequency R/T   186
E Application for Ship Licence and Certificate of
   Competency   187
F Documents to be carried   188
G Signalling: Rules 4 and 5 from the IYRU
   yacht racing rules   189
H United States inland rules: signalling
   differences   194
I Conversion factors   195
J Metric units and their symbols   196
K How to group digits and to use numbers with
   units of measure   198
L Signal lights   198
M Collision Regulations. Rule 3; Part C; Part D;
   Annex I, II and III (with 1983 amendments)   199
N Port Traffic Signals 1983   218

The main contents list gives a general guide, but to find more detailed information refer also to the chapter lists at the beginning of each part (1 to 3), where material in each chapter is listed.

Extracts from international regulations for preventing collisions at sea (1972, amended to 1983) and the 1981–84 international yacht racing rules are reproduced by kind permission of HM Stationery Office and the International Yacht Racing Union respectively.

Preface

Considering that man has been using boats of one sort or another for long-distance passage making since the time of the Phoenicians, it is strange to realize that, until very recently, he has always used much the same methods of signalling. Obviously the actual equipment became more efficient, but the basis has been the use of flags, shapes, sound and fixed or flashing lights.

Today, man still uses flags, shapes, sound and fixed or flashing lights, but he also uses radio and it is that – the use of radio – that has revolutionized the business of communicating at sea.

Modern man takes radio for granted and it is natural that he should because, even on ships, it has been in use since early this century; but the real revolution has been the introduction of a type of radio that is suitable for even the smallest boat. This has transformed the whole communication business – and all within the past decade.

It is reasonable to suppose that the Phoenician seaman used a flare-up light – created by dipping a piece of wick into a container of colza oil – perhaps three thousand years ago. A flare-up light produced by dipping cotton waste into a paraffin can was still used as a pilot signal until very recently, and 'flames on the vessel' is still recognized as an international distress signal. Nevertheless, modern man, when he wants to communicate, would not even think of trying to set light to cotton waste, even if he had any; he would turn to his radio.

There are still vitally important methods of signalling other than radio – pyrotechnics being the most obvious – but it is radio that matters most and it is radio that has priority in this book. Only a few years ago the idea of having instant communication with others, both ashore and afloat, via a 'magic' box no larger than a 'Pilot' book, would have seemed like science fiction. In a few years' time when nearly all traffic movements in commercial ports, all tidal information, all weather forecasts and almost all the instructions to ships from their headquarters will come over the air the mariner will find it difficult to remember what it was like before VHF R/T became the norm.

Yachtsmen go to sea for pleasure and they do it, as often as not, to get away from the controls and pressures that regulate their working lives. I hope that this book will be able to show that the ability to signal, to communicate with others, adds to the pleasure that is to be had from going to sea. Yachtsmen are rarely *required* to communicate – as ships are for environmental or port operational reasons – but small craft are a part of the maritime scene and the greater the congestion, the greater will be the need to be able to make a signal.

That signal might be as straightforward and unambiguous as the message conveyed by an anchor ball – 'I am at anchor' – or it might be as complicated and important as the relay of vital messages during a search and rescue incident with several stations involved.

Despite all the political or religious difficulties that flare up and then die down again in different parts of the world, seamen learned many decades ago that ships need to communicate, and to do so in a universally recognized manner. Everyone uses the same disciplines and the same procedures and, when necessary, the same codes.

Bernard Hayman
Burnham-on-Crouch
January 1983

PART 1 Radio Telephony

Introduction 3

Chapter 1: VHF R/T is really a giant party line 7
The 'capture effect' — How many channels? — Simplex and duplex — Public correspondence — Calling — Intership — Distress — Urgency — Safety — Port operation — Dual-watch — Power (output) — Power (input) — Private channels — HM Coastguard — Safety traffic.

Chapter 2: Rules and regulations 22
Books to be carried on board — Call sign — Ship Licence — Control of communications — Interference — Maximum intelligibility — Standard Phonetic Alphabet — Phonetic figures — Radio log.

Chapter 3: Standard Marine Navigational Vocabulary 33
Standard verbs — Responses — Urgent messages — Miscellaneous phrases — Repetition — Position — Courses — Bearings — Relative bearings — Distances — Speed — Time — Glossary — Procedural (Pro) words.

Chapter 4: Summary of recommended techniques and procedures 41
VHF Technique: Preparation — Repetition — Power reduction — Communications with shore stations — Communications with other ships — Distress — Calling — Watchkeeping.
VHF Procedure: Calling — Calling an unknown ship — Acknowledgement.

Chapter 5: Calling procedures 45
Intership — Port operation — Channel M — Public correspondence — Calling a CRS in the UK — The UK exception — Calling coast radio stations outside UK waters — Duration of calls and time limits — CRS calls to ships: traffic lists, direct calls, selective calling — Distress and urgency calls.

Chapter 6: Distress and urgency procedure 58
Carrier waves — Distress — Acknowledgement of distress — Mayday relay — Distress traffic — Urgency calls — Safety calls — Control of distress traffic — Imposing silence.

Chapter 7: Jargon 69
Useful abbreviation — Confusing abbreviation — Verbosity — The misuse of procedural words.

Chapter 8: HM Coastguard and communications 73
Channel Zero, 67 and 73 — Precedence indicators — Intership working during SAR — Auxiliary Coastguards — Listening watch: cruising.

Introduction
In the beginning
Although radio has been in use since early this century, it did not begin to have any worthwhile effect on communication at sea until the late 1920s; 70 years ago fewer than 300 UK ships had radio. Today there is a wide choice of equipment that is both cheap and reliable, and suitable for even the smallest yacht.

Radio is now compulsory for all but the smallest commercial craft and, within the next few years, it will be considered to be as much a part of the equipment of the coastal cruising yacht as her anchor and chain: it will be almost unthinkable (and for certain ports, illegal) to go to sea without one.

Without doubt the principal reason for this revolution has been the introduction of what is known as VHF (very high frequency) radio.

The appendices and tabulated frequency tables on pages 10–11 give some of the technicalities, but only *some* because the book is not a treatise on the subject of radio. It is designed to help the yachtsman to communicate at sea. The equipment he uses is very simple indeed. It is comparable to the minimal complexity of, say, an office dictating machine. However, although it is not really necessary to show anything at all about how a dictating machine works, it does help to understand where the maritime frequencies fit into the overall spectrum and how VHF differs from the other maritime bands.

Telephony and telegraphy

If all forms of maritime communication are taken into account, these now include many new methods, such as telex and satellite communication; but as far as the basic exchange of ideas – conversation – is concerned there are two different types: telegraphy (abbreviated as W/T, which stands for wireless telegraphy) where the telegraphist uses a telegraphic key, and telephony (abbreviated as R/T, for radio telephony) where the means of communication is speech. The French use the expression TSF for radio (wireless) and it, too, translates as 'telegraphy without wires'.

In the overall frequency grouping, W/T is used, as it has been since the 1920s, for Morse traffic, and on what are called high frequencies (HF). Given the right conditions and the right type of aerial installation, HF communication is possible over immense distances but, from the small craft point of view, there are several difficulties. The equipment is comparatively bulky and heavy. It consumes a great deal of electricity; and Morse needs a professional operator to achieve any worthwhile speed.

High frequency communication, developed during the 1930s, was followed by medium frequency (MF). Most national broadcasting as well as most medium-range maritime communication is on medium frequencies. For the purposes of this simplified guide to the spectrum, it is probably easiest to try to forget all about 'metres' as a means of measuring frequency and refer only to Hertz. One Hertz is one cycle per second. BBC 4, broadcasting on 1500 metres, is now referred to as being on 200 kHz (200 000 hertz). All but the oldest maritime radio is now marked in Hertz and even if 'pop' radio enthusiasts like to put 'stickers' on their car windows advertising their favourite local station as 'so many metres', I shall not refer to metres again. (The scientific world has adopted the Hertz as a unit in recognition of the German scientist Heinrich Hertz, who experimented as long ago as 1887 in electric waves: hence the capital letter in the abbreviated form Hz.)

Although radio has had such a tremendous impact on world affairs, it was for shipping that many early experimenters saw the greatest scope. Among the foremost pioneers was the Marconi Company and in view of the recent revolution in the use of R/T for small craft, I cannot resist the diversion of quoting from a memorandum from that company to Sir William Preece of the Post Office. It is dated 1907 and reads:

> "In the existing state of the art an expert telegraph operator must be carried on every ship. This will however be changed within a period of time measurable in months. Wireless telephony working with apparatus of great simplicity will take the place of the existing plant: word of mouth will supersede the Morse alphabet; the skilled operator will no longer be required; a ship's officer after undergoing a very short training will be able to adjust the apparatus and send and receive messages with much greater ease than flag and lamp signals can now be used. The lower cost of the telephonic installation will lead to its adoption on the meanest tramp steamer and in sailing ships. We are, in fact, quite close to the time when wireless communication will supersede flag and other systems of visual signalling in the Mercantile Marine of the World."

For 'a period of time measurable in months' read 60 years but otherwise the writer of that memorandum obviously realized the impact that the ability to communicate by radio was likely to have.

What frequency?

The normal home radio receiver is likely to have at least three 'bands', which are usually called long, medium and short; this is not only too simplistic a summary to describe maritime sets, but unlikely to be accurate. A receiver on a boat will have long wave, to receive essential stations such as BBC Radio 4 on 200 000 Hz (200 kHz), and also medium frequency, but here the direct comparison with traditional home radio ends.

The medium frequency band extends from 300 to 3000 kHz and includes most of the land-based broadcasting stations as well as the frequencies used by the coast radio stations throughout the world. This band has a range of, say, 200 miles – depending on aerials and power output – but it is increased at night and has the problems of distortion, on occasion, with which everyone is familiar.

For greater distances, high frequency (HF) sets are carried by larger ships and these allow communication of many hundreds and sometimes thousands of miles. However, very high frequency (VHF) radio waves behave in a different manner from HF, despite the fact that the HF and VHF bands are next to each other in the spectrum. HF extends from 3 to 30 MHz (3000 to 30 000 kHz), and the VHF from 30 to 300 MHz, but within that wide band – wide in the frequency sense – the maritime VHF channels all fall between 156.0 and 174.0 MHz. Furthermore, the channels normally licensed for small craft use are even narrower; from 156.025 to 162.025 MHz.

The tremendous difference between HF and VHF is that whereas the former is for use over very great distances, VHF is confined – or nearly so – to line-of-sight communication.

In Appendix D (page 186) there is a brief discussion of MF radio equipment. It includes something about the new breed of MF equipment that has only recently become legal for use on those UK ships where installation of MF sets is voluntary. During 1981, and following discussions between the Royal Yachting Association and the Home Office, the type-approval regulations for equipment for yachts that wished to carry MF radiotelephones were relaxed. As a result it became appreciably cheaper to fit MF equipment in small craft. However, despite this relaxation, and despite the introduction of Citizens' Band radio in the UK (which also became legal during 1981), it is the VHF frequencies that account for practically all small craft communication. It is VHF R/T that has caused the 'revolution'.

1 VHF R/T is really a giant party line

Nowadays the ordinary house telephone is so much a part of everyday living that it is taken for granted. As soon as a child begins to talk it will be encouraged to say 'Hallo, Granny', and by the time it is, say, ten it will use the telephone to speak to friends or to take messages for its parents.

When using radio to make what is called a 'link-call' – when the caller is connected to the person to whom he wants to speak via the land-line telephone service – the radio link (with one very important exception which will be explained) is exactly like using the instrument at home, but all other uses of R/T on a boat are different: in many respects, it is unfortunate that the device was ever called a telephone at all.

R/T – and particularly VHF R/T – is really a gigantic 'party line', with hundreds listening but only one speaking at one time. Once that point has been appreciated a gread deal of the confusion and wasted time on the air will be avoided. The point is a simple one.

If a man picks up a telephone, on a party line, and starts to speak into it when it is already in use, there is confusion and 'crossed lines'. The same thing applies with VHF radio, except that with a land-line telephone the fact that the line is in use is obvious to anyone willing to wait a few seconds and to listen. At sea, on the other hand, the use of a frequency is not merely a matter of listening for a moment. Even with a good sense of discipline there is some confusion. Without discipline there can be chaos.

The one very important exception referred to above is a phenomenon called the 'capture effect' and it is not until this is understood that the confusion can be minimized.

The 'capture effect'

Regarding VHF R/T as a gigantic party line is correct in the sense that a great many people are sharing the same means of communication. The expression 'party line', however, is not completely accurate. On a land-based party line there are several subscribers sharing the same telephone wire. If one subscriber starts to speak without first pausing and listening to see if the line is in use, his voice is merely superimposed on conversation in progress. With VHF radio, on the other hand, the loudest signal 'captures' the air waves and blots out all other traffic.

With the longer-range signals on MF R/T there is also the problem of one station interfering with another, but with MF the signal is usually being bounced off layers of air in the upper atmosphere. However, with VHF – and also with ultra high frequency (UHF) – the signals are approximately line-of-sight and the diagram in figure 1 shows what happens.

This 'capture effect' phenomenon explains why it is so common to overhear half conversations. If you are near C in the figure you will hear

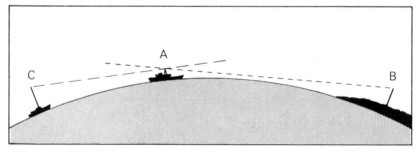

Figure 1 *VHF R/T has a range which is approximately line-of-sight. However, because the strongest (nearest) signal 'captures' the frequency being used a transmission from "C" would blot-out one that "A" was receiving from "B".*

The fact that "C" might be calling yet another station and the fact that "C" cannot hear anything of "B"'s transmission, makes no difference. Everyone must wait and listen before transmitting; if they do not both their own and everyone elses traffic is likely to take twice as long.

everything that A is saying to B but, like C, you will hear nothing of the replies. Thus, not only must you wait, but you must wait long enough to ensure that the frequency is properly clear of traffic if interruption is to be avoided.

How many channels?

During the 1970s, when VHF radiotelephony began to be used in small craft, the equipment available often had only 10 or 12 possible channels – and the portable sets merely 3 or 5. In those days there was a choice of a total of 28 channels.

Today, nearly all sets other than portables have all the 55 channels that are available, and most of them also have the facility to add additional, so-called private channels. Therefore the question of which channel to install in any particular set is now not often relevant, but it is still essential to understand how these 55 channels are allocated, and which can be used for what purpose.

Figure 2 shows the manner in which the principal channels are allocated. The apparently illogical numbering in the first column is a result of the interleaving of channels 60 to 88 in the previous group of 01 to 28. That was done in 1972 when the technical development in the equipment being used allowed the widths of the channels to be halved. Technically, it would be possible to halve the existing channels yet again if the equipment stability is good enough, but there are no signs that this will be permitted in the immediate future.

Later, in the chapters devoted to procedure, there will be a great deal to say about how particular situations and facts are presented on R/T, but note that it is not mere bureaucratic gobbledegook but good sense to write '01' rather than just '1'. Since most of the channels have two-figure numbers (10 and above), there is logic in writing numbers below 10 as two figures. Furthermore, the professional communicator will use this two-figure notation in transmitted speech: it is channel 'six seven', not 'sixty-seven'.

Note that although there are four groups of columns in figure 2 – Intership, Port Operation, Ship Movements, and Public Correspondence – the Port Operation and Ship Movements columns are so nearly the same – the only difference being in the use of Ch. 15 and 17 – that they will be treated as one at this stage.

Transmitting Frequencies in the 156–174 MHz Band.

FOR STATIONS IN THE MARITIME MOBILE SERVICE

Channel designators		Notes	Transmitting frequencies (MHz)		Inter-ship	Port operations		Ship movement		Public correspondence
			Ship stations	Coast stations		Single frequency	Two frequency	Single frequency	Two frequency	
	60	j)	156.025	160.625			17		9	25
01		i)	156.050	160.650			10		15	8
	61		156.075	160.675			23		3	19
02			156.100	160.700			8		17	10
	62		156.125	160.725			20		6	22
03		i)	156.150	160.750			9		16	9
	63	i)	156.175	160.775			18		8	24
04			156.200	160.800			11		14	7
	64		156.225	160.825			22		4	20
05			156.250	160.850			6		19	12
	65		156.275	160.875			21		5	21
06		h)	156.300		1					
	66		156.325	160.925			19		7	23
07			156.350	160.950			7		18	11
	67	n)	156.375	156.375	10	10		9		
08			156.400		2					
	68	p)	156.425	156.425		6		2		
09		o)	156.450	156.450	5	5		12		
	69	p)	156.475	156.475	9	11		4		
10		n)	156.500	156.500	3	9		10		
	70	o)	156.525		6					
11		p)	156.550	156.550		3		1		
	71	p)	156.575	156.575		7		6		
12		p)	156.600	156.600		1		3		
	72	o)	156.625		7					
13		p)	156.650	156.650	4	4		5		
	73	n)	156.675	156.675	8	12		11		
14		p)	156.700	156.700		2		7		
	74	p)	156.725	156.725		8		8		
15		g)ll)	156.750	156.750	12	14				
	75	m)			Guard-band 156.7625–156.7875 MHz.					
16			156.800	156.800	DISTRESS Safety and Calling					
	76	m)			Guard-band 156.8125–156.8375 MHz.					
17		g)ll)	156.850	156.850	13	13				
	77		156.875		11					
18		f)	156.900	161.500			3		22	
	78		156.925	161.525			12		13	27
19		f)	156.950	161.550			4		21	
	79	f)p)	156.975	161.575			14		1	
20		f)	157.000	161.600			1		23	
	80	f)p)	157.025	161.625			16		2	
21		f)i)	157.050	156.050 or 161.650			5		20	
	81		157.075	161.675			15		10	28
22		f)	157.100	161.700			2		24	
	82		157.125	161.725			13		11	26
23		i)	157.150	156.150 or 161.750						5
	83	i)	157.175	156.175 or 161.775						16
24			157.200	161.800						4
	84		157.225	161.825			24		12	13
25			157.250	161.850						3
	85		157.275	161.875						17
26			157.300	161.900						1
	86	q)	157.325	161.925						15
27			157.350	161.950						2
	87		157.375	161.975						14
28			157.400	162.000						6
	88	j)	157.425	162.025						18

Taken from *Admiralty List of Radio Signals Vol 6*

(a) The figures in the column headed 'Intership' indicate the normal sequence in which channels should be taken into use by mobile stations.

(b) The figures in the columns headed 'Port operations', 'Ship movement', and 'Public correspondence' indicate the normal sequence in which channels should be taken into use by each coast station. However, in some cases, it may be necessary to omit channels in order to avoid harmful interference between the services of neighbouring coast stations.

(c) Administrations may designate frequencies in the intership, port operations and ship movement services for use by light aircraft and helicopters to communicate with ships or participating coast stations in predominantly maritime support operations under the conditions specified in the Radio Regulations. However, the use of the channels which are shared with public correspondence shall be subject to prior agreement between interested and affected administrations.

(d) The listed channels with the exception of 06, 15, 16, 17, 75 and 76, may also be used for high-speed data and facsimile transmissions, subject to special arrangement between interested and affected administrations.

(e) Except in the United States of America, the listed channels, preferably two adjacent channels from the series 87, 28, 88, with the exception of 06, 15, 16, 17, 75 and 76, may be used for narrow-band direct-printing telegraphy and data transmission, subject to special arrangement between interested and affected administrations.

(f) The two-frequency channels for port operations (18, 19, 20, 21, 22, 79 and 80) may be used for public correspondence, subject to special arrangement between interested and affected administrations.

(g) Until 1 January 1983, the effective radiated power of ship stations on channels 15 and 17 shall not exceed 1 W.

(h) The frequency 156.300 MHz (channel 06) may also be used for communication between ship stations and aircraft stations engaged in coordinated search and rescue operations. Ship stations shall avoid harmful interference to such commmunications on channel 06 as well as to communications between aircraft stations, ice-breakers and assisted ships during ice seasons.

(i) In France and in Belgium, the frequencies 156.050, 156.150 and 156.175 MHz are used as ship station frequencies in channels 01, 03 and 63 respectively and as coast station frequencies in channels 21, 23 and 83 respectively when the latter are used in the special semi-duplex public correspondence systems employed with 1 MHz separation between transmit and receive frequencies. These special provisions will cease to be used not later than 1 January 1983.

(j) Channels 60 and 88 can be used subject to special arrangements between interested and affected administrations.

(k) The frequencies in this Table may also be used for radiocommunications on inland waterways in accordance with the conditions specified in the Radio Regulations.

(l) Channels 15 and 17 may also be used for on-board communications provided the effective radiated power does not exceed 1 W, and subject to the national regulations of the administration concerned when these channels are used in its territorial waters.

(m) This guard-band will apply after 1 January 1983.

(n) Within the European Maritime area and in Canada these frequencies (channels 10, 67, 73) may also be used, if so required, by the individual administrations concerned, for communication between ship stations, aircraft stations and participating land stations engaged in coordinated search and rescue and anti-pollution operations in local areas, under the conditions specified in the Radio Regulations.

(o) The preferred first three frequencies for the purpose indicated in Note c) are 156.450 MHz (channel 09), 156.525 MHz (channel 70) and 156.625 MHz (channel 72).

(p) These channels (68, 69, 11, 71, 12, 13, 14, 74, 79, 80) are the preferred channels for the ship movement service. They may, however, be assigned to the port operations service until required for the ship movement service if this should prove to be necessary in any specific area.

(q) This channel (86) may be used as a calling channel if such a channel is required in an automatic radiotelephone system when such a system is recommended by the CCIR.

Figure 2

The word 'intership' is self-explanatory, as is the phrase 'port operation', but 'public correspondence' may not be so obvious. It is the expression used for ship-to-shore traffic via a coast radio station, whereby a telephone call is made from a ship and the caller is then connected to any land-line telephone system. In practice the service is usually referred to as a 'link-call'.

Note that some channels are reserved for one purpose only – Ch. 06 for example is solely for intership traffic – and some serve more than one category of traffic. Note, too, that some are marked 'single frequency' and others 'two frequency'; the difference is fundamental to an understanding of the allocation of channels.

Simplex and duplex

Channels that have only one frequency allocated to them – Ch. 06 for example – are called simplex channels. Both transmission and reception are on the same frequency but not, as on a land-line telephone conversation when both can speak at the same time. On a land-line telephone there is confusion if both speak at once, just as there can be if two people both speak at the same moment when face to face, but the problem is merely a risk of not understanding. If anyone tape-recorded the conversation the tape would pick up both 'sides'.

With a simplex channel on VHF R/T the transmission and reception circuits in the radio are different and if both people start to speak at once neither will hear anything at all of the other's conversation; neither receive circuits will be alive. The transmission circuit is activated by a 'press-to-speak' switch in the handset and is invariably spring-loaded so that, in theory at any rate, it will never be left 'on' by mistake. In other words a VHF R/T is always in the receive mode, when first switched on, and it remains on receive unless needed for actual transmission.

If something goes wrong, and the press-to-speak button is jammed in the 'on' position, the set will transmit what is called a 'carrier wave', even when no one is speaking. That 'carrier' blocks the frequency in the same manner as would occur if it was actually in use, and the fact that the set is now switched to the transmit mode prevents reception anyway. The full implication of this state of affairs will become clearer later, but, for the moment, the point to note is that a simplex channel allows one-way traffic only: press to speak and release to receive.

Duplex operation allows for simultaneous transmission in both directions, but not only are two different frequencies used, there have to be two aerials (or a special duplex filter) as well. With full duplex a two-way conversation similar to that on a normal land-line telephone is possible. However, full duplex is considerably more expensive both to buy and to install, and it is not often found on small craft.

Semi-duplex is a compromise that allows duplex channels to be used on what is basically simplex equipment. It has simplex at one end of the conversation and duplex at the other. Note that all the public correspondence channels are duplex, but they can be used with semi-duplex equipment because, although the small craft (simplex) end of the exchange still has to press to speak, as with ordinary simplex channels, the switching at the coast radio station end is automatic.

In practice only the simplest sets – mostly portable – are simplex only, because simplex equipment can never be used on duplex channels. Therefore it cannot be used for public correspondence. As full duplex costs considerably more, the vast majority of all small craft VHF R/T sets are semi-duplex and, unless stated otherwise, all the procedural points that follow will be applicable to that mode. As far as procedure at the small craft end of any traffic is concerned, simplex and semi-duplex are the same: it is the full duplex equipment that is different.

Public correspondence

As already mentioned, some channels have two frequencies, others one. Also some have two purposes and others one. However, figure 2 shows that all the channels allocated for public correspondence are ship-to-shore channels in that they are either public correspondence alone, or they are shared with port operations; all are two-frequency channels.

With two-frequency channels the ship transmits on one and receives on the other and the coast station does the opposite. Thus if a ship switched to say Ch. 25 – a public correspondence working channel – while awaiting her turn to be called by the coast station, she might hear half of any conversation that was taking place at the time. Listening on Ch. 25 her set would be alive on the receive side on 161.850 MHz. The set would be switched off that frequency and switched to 157.250 MHz only when she used her press-to-speak switch to transmit to the coast station. In practice, however, because of what the professionals call 'splash-over' either at the ship or at the coast station, the third party, awaiting his turn, often hears both sides of the other conversation: the side of the traffic using 'his' transmit frequency is being re-broadcast on the receive frequency so that he hears both. This ought not to happen. However, it does and it merely underlines the basic point that VHF R/T is a party line. Dozens may be listening even to what might be thought to be private conversations.

Calling

With a shore-based telephone, a call is made by dialling the required number and this alerts the subscriber with a bell. If the line to the exchange is in use, the equipment automatically selects another, and so on, to the point when, at a busy time of day the subscriber may be told that all lines are busy.

With 55 different channels from which to choose on VHF R/T, a beginner in the art of wireless communication might assume that there are 55 different ways of calling another station, but that is far from the truth. There is only one channel, in normal circumstances, that actually works for calling: the channel to which the receiving set is tuned.

Intership

It should be self-evident that if one station is tuned to channel A and another to channel B, a call on channel C will not be heard by either. A 55-channel set can switch to 54 different channels – the choice depending on the nature of the traffic – once communication has been established on one, and the one that is usually used for intership calling is Ch. 16.

If it were possible to assume that merely a handful of ships might be listening at any one time, the volume of traffic on any designated calling frequency would be slight. In fact the regulations in force today still refer to the need to keep a call on Ch. 16 to not more than one minute. In practice, however, there are many areas where there might be hundreds of ships within range of each other and the pressure on Ch. 16 makes such demands on the possible air time that a call lasting more than a few seconds can be a nuisance to others.

Later, when procedure is discussed in greater detail in Chapter 5, the precise form of words for a call will be examined, but note that a 'calling' channel is not intended for passing messages – other than distress; the object is solely to establish radio contact for long enough to agree on an appropriate working channel. Once the procedures have been understood, that need rarely take longer than a total of 12 to 15 words.

Note, too, that Ch. 16 is designated as the distress and safety, as well as the calling frequency. It must therefore be obvious that while the air waves are engaged with calling traffic they are not available for distress or safety messages. A distress message transmitted while another station in the vicinity is using the channel for a call, is merely wasted effort. There is a good chance that no one at all will hear it.

Distress

In the language of the sea, distress has a reasonably precise meaning. The use of a distress signal – and there are several – means: 'A ship . . . is threatened by grave and imminent danger and requires immediate assistance.'

A full list of all the internationally agreed distress signals is included in figure 22 on page 159, but from the V H F point of view the spoken word 'Mayday' precedes any distress message. By international agreement a distress call has a recognized form which is described in detail in Chapter 6, but for VHF R/T (where at present there is no automatic alarm signal to precede the distress call, as there is for MF and HF radio) a distress call begins with the word 'Mayday' repeated three times.

Distress transmissions are always on Ch. 16.

Urgency

Urgency traffic has priority over all other traffic, except distress, and is for use when a ship has: 'A very urgent message to transmit concerning the safety of a ship or the safety of a person.' An urgency call is preceded by the signal 'Pan-Pan', repeated three times, and urgency traffic is also transmitted on Ch. 16, although there are circumstances when a long message might be transferred to a working channel.

Unfortunately, the usefulness of having a distinction between distress and urgency is not often understood by small craft. However, there are degrees of urgency, just as there are degrees of almost any other state. One of the differences between 'requiring immediate assistance because the ship is in imminent danger' and 'requiring assistance, because of concern for the safety of the ship or of a person', lies in the effect on the legal responsibilities of those involved.

Safety

Ch. 16 is designated the 'Distress, Safety and Calling' channel and the third degree of priority after distress and urgency is given to safety traffic. This has priority, on Ch. 16, over all other traffic except distress or urgency, and here the call is preceded by the French word 'Sécurité' (pronounced in the anglicized fashion SAY-CURE-E-TAY), spoken three times

As with the other priority calls, safety traffic starts with a call on Ch. 16, but the message itself is normally passed on a working channel. Safety messages might concern a buoy out of position or some unusual tow to which a coast radio station wanted to draw the attention of shipping.

Port Operation

Although many of the large ports now rely heavily on VHF R/T for almost all their operations, this idea is comparatively recent. The advantage to all parties of a continuous means of communicating first became obvious in the large ports like London and Southampton, and as the idea of a continuous port operation watch became established, so did the principle of asking shipping to call on a working frequency.

Small ports, often with an office manned for short periods only, still use Ch. 16 as the calling channel but increasingly there is a move away from that traditional method.

The principle is simple. All ships that are not required by law to monitor MF frequencies are asked to monitor Ch. 16 at all times when they are at sea, because of the obvious advantages of having one channel open for distress and calling. However, the more that channel is used for calling, the less it is available for any of the priority traffic that has just been outlined. Thus the ports that have the equipment to allow them to do so monitor Ch. 16 *and* their own port operation channel. When a ship knows that a particular port will be monitoring its port operation channel she calls on that, and speaks directly to the port, and does not need to use Ch. 16 at all.

Today, the official advice to everyone is to call on a working channel whenever it is known to be watched. The great advantage of following this advice is that it relieves Ch. 16 for traffic that it has to accept (such as priority traffic) and for calls from other ships for intership calling.

If a port is large enough to have more than one operator on duty at all times and more than one radio, it is feasible for calls to be accepted on Ch. 16 and on a working channel at the same time. In practice only the largest ports have that facility and the compromise is to use what is called dual-watch: the facility to listen to two different channels at the same time.

Dual-watch

Only a few years ago a dual-watch facility was something that only a few sets were able to offer – as an optional extra. Today most multi-channel sets have dual-watch built in as standard and there are occasions when the facility is very useful: entering a commercial port is one of the most obvious.

What happens is that the set switches itself back and forward between two channels, several times every second, and then 'locks on' to either channel as soon as there is a signal to receive. Unless special arrangements are made, the equipment will 'scan' between Ch. 16 and whatever channel the set is switched to. The choice is therefore open for dual-watch with Ch. 16 and any working channel.

The clever point regarding this high speed scan is that Ch. 16 is given priority: if a signal comes on Ch. 16 the set will switch to that, even if it was receiving traffic on the working channel at the previous moment.

The principle is eminently sensible. With dual-watch switched on, a man can monitor, say, his local port operation channel – to give him information about local movements, weather and any local hazards – and, at the same time he can monitor Ch. 16 for safety reasons, and in case any other station wishes to call him.

In practice the idea does not work too well if there is a great deal of traffic in the vicinity because any message on the port frequency is likely to be interrupted by some other ship calling a CRS or yet another ship on Ch. 16.

Once a port frequency has begun to pass a message that might be important – for example, a special local visibility warning – it is always possible to switch off the dual-watch scan (to ensure the working channel traffic is not interrupted) and then switch back again to the dual-watch mode. But that tends to be a nuisance and the problem merely underlines (yet again) the need for discipline on Ch. 16.

Power (output)

In view of what has been said about the capture effect and the line-of-sight range of VHF, it might be assumed that everyone would be tempted to install more and more powerful sets. However, and for that very reason, everyone is limited to a maximum of 25 W from the set and the only way to improve 'your' output – compared with 'his' – is to ensure that the aerial is unobstructed (in the radio sense) and that the special cable between set and aerial is of good quality. You should use a low-loss co-axial cable and take a great deal of trouble to ensure that the external cover does not become damaged. Once water has seeped into a co-axial lead-in, the power actually transmitted falls off to an alarming degree.

A hand-held portable will probably not have an output of more than 1 W, compared with the usual 25 W for the permanently installed set working off the ship's electrical supply, but all sets are required to have a 'low-power' switch, which limits output to 1 W. It is recommended that this should be used whenever possible to minimize the interference that any ship's transmissions cause.

Power (input)

The power supply necessary for a VHF R/T is small enough to be disregarded for any well-found cruising boat that already has an efficient electrical wiring circuit and a healthy means of charging the batteries. On receive only – the state in which the radio will be used for the greater part of the day – a typical set consumes about 0.8 A on a 12-volt circuit. That is less than the current from one navigation light bulb. In other words, a VHF R/T can be left on all day and all night for a battery drain of about 20 ampere/hours – perhaps half an hour charging time from an alternator.

On transmit, however, the consumption may be as high as 8 A when on full power, but as that might be for only a few minutes in the day the battery drain is usually acceptable. Furthermore, it compares favourably with the drain for MF and HF equipment, which can be 20 or 30 A or more.

On a low power of 1 W the transmit drain on the battery is considerably less – probably only about 1 A. It would not be practical to take more than this from a small rechargeable cell used for the hand-held portable, and somewhere between those extremes the type of set found on, say, an RNLI inshore lifeboat might have an output of 5 W, which is adequate for her comparatively limited range of operation.

Private channels

In addition to the three main groups of channels already mentioned – intership, port operation and public correspondence – which are listed in figure 2 – there is another large group of VHF channels allocated to what are normally referred to as private purposes. A salvage concern or an oil company might have a VHF channel allocated for use by its ships to communicate to the company's base station but, in general, private channels of this kind are not of interest to the small craft skipper and he is not licensed to use them.

One exception is that for UK small craft 157.85 MHz can be licensed for use by marina operators and yacht clubs as a simplex channel. They may employ it for marina business or for yacht race management. Nowadays, when a Ship Licence is issued, the permission to use 157.85 is granted in the form of a letter at the same time, but a marina or club needs a special 'base station' licence, which is not cheap.

Licensing arrangements generally are discussed later, but it should be noted that 157.85 – referred to as Ch. M – is not an intership channel. Unfortunately, however, it is sometimes used as such. It is a yachtsman's private port operation and race control frequency and it is illegal to use it for any other purpose.

There are no other special channels normally used by UK licence holders, but in the United States several special rules apply. They are summarized in Appendix B, along with French and Belgian special frequency allocations although in 1983 in these latter countries, the anomalies are supposed to have been phased out.

HM Coastguard

The only other channel regularly used by small craft, but only by a specialized few who have to be separately licensed by the Home Office, is Ch. 00 (Zero) 156.00 MHz. It is employed primarily by HM Coastguard as a private channel and is also the channel used by the Royal National Lifeboat Institution, by certain police craft, and by a few Auxiliary Coastguards. Coastguard vehicles – licensed as mobile stations – rescue helicopters, and a few other specialized ships also use Ch. 00. Only the professionals and a few specially licensed amateurs under the direct control of HM Coastguard, have the use of this vital line of communication for emergency traffic. The role of the Auxiliary Coastguard is discussed in Chapter 8.

Safety traffic

Channel 67 has one 'special' use, as far as small craft are concerned, in that the British authorities allow its use for direct communication between ship and HM Coastguard stations. Ch. 67 is not a public correspondence channel in the normal sense in that it cannot be used for passing messages to third parties, but its use for traffic to HMCG was authorized in the late 1970s for what is officially termed 'safety traffic'.

This concession, which applies in the United Kingdom only, is a considerable help and it reflects the interest that the government department concerned – the Marine Division of the Department of Trade – has in safety matters. In the UK we now have an almost continuous coverage of coastal waters to between 25 and 40 miles offshore, not only from the British Telecom coast radio stations – British Telecom (until 1981 the Post Office) being the body responsible for all ship-to-shore traffic for which there is a charge – but with HM Coastguard radio stations as well. The only gaps in the Coastguard cover are a few parts of the more remote Scottish coastline.

Until the late 1970s the Post Office (as it was then) had the responsibility for relaying all distress and safety traffic. In 1976 HM Coastguard, which was already charged with coordinating search and rescue operations, assumed responsibility for keeping a constant distress watch in Ch. 16 and subsequently, in 1979, the idea of a safety channel was introduced.

Ch. 67 is not a distress channel, in the normal sense, even if it is sometimes used for distress working. Internationally speaking it is still an intership or port operation channel. In Britain its prime function is as a channel that can be used for direct communication between small craft and HM Coastguard 'on matters affecting the safety of the vessel'.

It is a working channel and not one that is normally monitored. Small craft therefore call on Ch. 16 – the distress, safety and calling channel – and state they have 'safety traffic'. They will then be asked to change to Ch. 67. In practice the official phrase 'matters affecting the safety of the vessel' is generously interpreted. For example, members of the Coastguard are not weather forecasters – that is not their job – but they will normally give yachtsmen an indication of present weather in their vicinity and will also be able to give the latest shipping forecast for the local sea area. What the service is not allowed to do is to pass messages to others: R/T traffic to third parties (other than during distress) goes via coast radio stations and the British Telecom land-line.

2 Rules and Regulations

Nowadays, with almost instant communication via a land-line telephone with most parts of the world, there is an obvious tendency for those unfamiliar with the complexities of the subject to believe that there are no rules affecting VHF R/T that really matter; and that where there are rules, they are the efforts of some faceless bureaucrat with little better to do than compose regulations to justify his existence. To a very small extent this is true. The regulatory side of any new development tends to drag behind the actual practice, and VHF R/T is no exception.

Historically, the wireless operator on a ship was a specialist. He was trained to operate the complex equipment and to use Morse at what, to the uninitiated, appeared breathtaking speeds. However, the equipment has been progressively simplified to the point where VHF R/T is installed on the ship's bridge, not in the wireless room, and, on small craft at any rate, there is a tendency for anyone to use it.

The precise wording of the law concerning this point is slightly contradictory. The *Handbook for Radio Operators* (para. 137) refers to the maritime bands between 156 and 162 MHz (the VHF R/T bands) being under the control of an operator holding the appropriate certificate of competence and goes on to state that 'Provided that the installation is under the control of such a qualified operator, other persons may use the radiotelephone service.'

In the actual Ship Licence, however, there is a clause which states that the station shall be used only by the holder of the appropriate certificate of competence 'or in the presence of and under the supervision of a person so authorized'. That phrase 'in the presence of' is obviously more demanding than the more general expression 'under the control of'. The authorities obviously mean the former to apply because the letter that is sent out to accompany the licence specifically draws attention to the clause quoted.

As far as most small craft are concerned the 'operator holding the appropriate certificate' will be the holder of a Restricted Certificate of Competency in Radiotelephony (VHF only). There are three documents involved, the Certificate of Competency, the Authority to Operate and the Ship Licence. The Certificate of Competence depends on the result of an examination. The Authority to Operate is normally attached to the Certificate of Competence but is generally restricted to British subjects and citizens of the Irish Republic and it can be withdrawn at any time by the Home Office if there are sufficient grounds to do so. The Ship Licence, on the other hand, is a once-only application, but there is an annual fee (£17.50 in 1982) payable to the Home Office. The Ship Licence has to be returned if the vessel is sold but the allocated call sign remains with the vessel while she remains in United Kingdom Registry.

The Ship Licence contains details of the frequencies which may be used, the types of emission, and the conditions under which the station must be operated (see Appendix E).

The law requires *at least* one person on board to be the holder of a Certificate of Competence in Radiotelephony, but it is most strongly recommended that all should take the examination. It is simple and largely oral and, with VHF becoming so much a part of the normal sea-going scene, the confidence obtained by passing the exam is tremendous when compared with the small effort involved. The examination fee (£20 in 1982) might suggest that the examination is complex and designed for specialists or professionals. That is not so. The holder of a Certificate of Competence has the additional advantage that he can use other R/T sets without having to be 'under the control of' somebody else who holds one.

After lengthy discussions between the Royal Yachting Association and the Home Office it was arranged in 1983 that all who want to study for and take the Certificate of Competence in Radiotelephony can do so as a part of an appropriate RYA teaching programme, or by applying directly to the RYA, without having to involve Home Office personnel.

Quite apart from the need to encourage discipline on R/T and to teach it as a part of the normal process of learning 'the trade' it is expected that the RYA will be able to administer and invigilate the examination very much more cheaply than was the case before.

Books to be carried on board

The law also specifies what documents have to be carried on board and one of those is the *Handbook for Radio Operators*, published by Her Majesty's Stationery Office. It is also available from most small craft chandler's shops. As might be expected from an HMSO publication, it is authoritative and accurate. Unfortunately it is also extremely difficult to read because it was written by men already familar with radio procedures and the 'jargon' that the professional takes for granted. Also it was written before the introduction of the 'VHF only' certificate and the result is that large sections of the book are incomprehensible to the beginner.

Nothing in what follows conflicts with the instructions and advice given in the *Handbook for Radio Operators*, but I have tried to offer the facts in a slightly more digestible form. The *Handbook* is excellent for reference, once the skeleton of the subject has been understood, but it is almost useless as a beginner's primer.

Call sign

When the application is made for a Ship Licence – and one can be issued to any British ship whether she is officially registered or not – a call sign will be included. This call sign, which is normally composed of four letters (or one figure and three or four letters), stays with the ship even if she is sold or changes her name – unless, of course, she passes out of British ownership, because the Ship Licence would then automatically be revoked and the new owners would have to apply for a new licence from their own authorities.

Most wireless telegraphy regulations are international and are controlled by World Maritime Administrative Radio Conferences held at regular intervals, but the call sign is akin to a number plate and nationality letter on a motor car: it identifies the precise ship and its nationality.

Names of ships are still used in a call because they are easier to remember than jumbles of letters, but it is the call sign that is the key identifier if there is any confusion, and it is the call sign that is used by coast radio stations throughout the world to determine where telephone charges have to be sent.

Appendix F includes a full list of the books and other documents that a vessel has to carry but the Ship Licence, with its proof of the call sign, is obviously the most important because, when cruising away from home, other administrations have the right to demand to see it.

Ship Licence

The Ship Licence fills four pages of A4 paper with fine print and it tends to read like the Rules and Regulations posted at the entrance to a public park; never to be read by-laws about playing ball games or making a noise. No one normally reads the park by-laws (even if they are necessary to give the park keeper his authority when occasion demands), but the rules for a Ship Licence are important. Misbehaviour in a park may upset half a dozen people near by; misbehaviour on radio can upset hundreds and can even cause loss of life if, for example, a search and rescue mission is thwarted by a lack of the ability to communicate.

The following are forbidden:
1. Transmissions that have not been authorized by the Master or by the person responsible for the ship.
2. Transmissions that do not identify the ship's name or call sign.
3. Transmissions that are grossly offensive, indecent or obscene.
4. The use of transmitting equipment which does not meet the Home Office performance specification.
5. The transmission of false signals of distress or other misleading messages.
6. The transmission of messages for third parties on land, other than via a coast radio station.
7. To make known the contents of any message, or even the fact of its receipt, not authorized by the Licence. (In other words to make any use of traffic overheard but not intended for you.)
8. To broadcast messages (other than safety messages) without a reply being expected.
9. To broadcast music.
10. To use channels other than those authorized by the Licence.

11 To make unnecessary transmissions or superfluous signals.
12 Transmissions to other ships, when in port, except when using port operation frequencies and service, or for the purposes of safe navigation.

This list of a dozen prohibitions is not the full story. It is an infringement of the regulations to close down a radiotelephone, for example, if there is distress working in which the ship might be involved, or without clearing all traffic on hand with a coast radio station. But, as might be imagined, the regulations were framed for a far wider public than yachtsmen.

If those 12 rules were always obeyed the problems with VHF R/T would be virtually non-existent.

Control of communications

With ship-to-shore traffic it is the shore station that is in control. A CRS might already have a queue of traffic of which the calling station was unaware – either because of the capture effect or because the calls were on other frequencies – so it is obvious that a CRS must be in a position to say, in effect, 'go away!'. The question of priority traffic will be referred to later but, remember, the shore always controls.

For intership traffic – and again we are not concerned with distress or urgency – it is the ship *called* which has control. In the following chapters I shall have a good deal to say about procedure and the so-called pro words ('pro' standing for 'procedural' not 'professional'). Note, however, that already there is a certain amount of jargon – words not normally used in conversation – and some of it is not only inevitable but positively helpful.

In any exchange there are at least two parties involved. The technical term for each is 'station': a 'ship station' or a 'shore station', but in practice the expressions are usually shortened. Nevertheless, it is often important to know which of the two stations initiated the traffic. When channels have to be changed, for example, the procedural rules lay down which station suggests the choice of working channel. Therefore the first piece of jargon to absorb is that the yacht, ship or other station starting the traffic is referred to as the 'calling station' and the recipient becomes the 'station called'.

Observant readers will have noted that the important difference is in the position of the word 'station'. The words 'calling' and 'called' sound so similar that they are put at opposite ends of the two-word phrase.

Interference

The reference to the 'capture effect' in Chapter 1 will have explained one of the main causes of interference: when one station calls another but is unable to hear that the station called is already busy. The second main cause of interference is when a man simply turns his channel selection switch to the one he intends to use and starts to speak. If he is within range of one or both halves of an existing exchange of messages his transmission will blot out one or other of them. That is why the distinction between calling and working frequencies is so important, and why it is essential to listen and wait before transmitting. It is also mandatory to wait if told to by a coast radio station, but any station controlling an exchange will normally indicate how long the calling station is being asked to wait.

An operator is not being rude if he says merely: *Wait – Out*. It is the correct way to deal with some situations.

Maximum intelligibility

If reception is perfect, if there is no interference, if there is plenty of time and if neither operator is under stress, then the need for special procedures hardly arises. However, life at sea is rarely like that and the principal reason for following agreed procedures is that you are more likely to be understood when you do. VHF R/T is an emergency tool, as well as a business and social convenience, and even if instant comprehension is not of vital importance during a chat with a friend, it is obvious that there must be no ambiguity in distress.

We in the United Kingdom are most fortunate in that the world is steadily changing to the use of English as the language of the sea.

From the Russian Baltic coast to France, every European country broadcasts navigational warnings. They go out every four hours every day of the year and on a pre-arranged pattern but at times carefully staggered so that transmission A does not interfere with B. It is a system that has been in force since 1975. All the broadcasts go out first in English, followed by the language of the country.

There are many other examples – weather bulletins for example are also usually in English – but even in English an unfamiliar accent can make words difficult to understand and, in difficult conditions, even the clearest tones can become garbled.

The following factors are important:

Rhythm   Any phrase spoken in ordinary conversation has a natural rhythm which helps to make it intelligible. That rhythm must be preserved when a phrase is spoken on a radiotelephone. A message should be given in complete sentences or phrases that make sense and not word by word. Care must be taken to avoid hesitation and never to say 'er' or 'um' after a word.

Speed   The sending operator must speak steadily at medium speed. If he speaks too fast, his words may be received in an unintelligible jumble, and if he speaks too slowly he will waste time as well as exasperate the receiving operator. The speed of speech must be kept constant throughout and the less important words must not be hurried.

Volume   The operator will normally speak slightly louder than in ordinary conversation, but he must not shout. In ordinary conversation the important words are usually stressed while the less important ones are likely to be slurred. This must be avoided. Each word must be spoken with equal clarity and the voice must not fade away on the last word in a sentence. The mouth must be kept fairly close to the microphone; if the head is turned away when speaking – for example, to look at a chart – the volume of speech received at the other end will drop and words may be lost.

Pitch   High-pitched voices are transmitted more successfully than those of a lower pitch, which explains why women can make such good communicators. In normal conversation the voice is often allowed to drop in pitch on the last syllable of each word and the last word of each phrase. This too must be avoided.

Standard Phonetic Alphabet

Although there have been different versions of the phonetic alphabet in use on R/T in the past, the countries of the world have now agreed on a standard form and the somewhat odd pronunciations are carefully chosen to suit those who cannot easily say particular sounds. Note, too, that the emphasis on the words can be almost as important as the words themselves, if reception is really poor.

Letter	Phonetic word	Spoken as
A	Alfa	AL fah
B	Bravo	BRAH voh
C	Charlie	CHAR lee *or* SHAR lee
D	Delta	DELL tah
E	Echo	ECK oh
F	Foxtrot	FOKS trot
G	Golf	Golf
H	Hotel	hoh TELL
I	India	IN dee AH
J	Juliet	JEW lee ETT
K	Kilo	KEY loh
L	Lima	LEE mah
M	Mike	Mike
N	November	no VEM ber
O	Oscar	OSS cah
P	Papa	pap PAH
Q	Quebec	keh BECK
R	Romeo	ROW me oh
S	Sierra	see AIR rah
T	Tango	TANG go
U	Uniform	YOU nee form *or* OO nee form
V	Victor	VIK tah
W	Whiskey	WISS key
X	X-ray	ECKS ray
Y	Yankee	YANG key
Z	Zulu	ZOO loo

This alphabet is an essential part of the business of communicating on a radiotelephone all round the world. The same phonetic spellings and emphasis are now used by everybody.

The phonetic alphabet is always used for transmitting call signs: never M–A–B–A; always *Mike Alfa Bravo Alfa*. It soon becomes second nature to say it like that.

What is not so easy, however, is to interpret the phonetic alphabet when it is used for longer words or groups of words. A call sign is only four or five letters or digits and letters, and the mind can easily retain what was said. For a two- or three-word name the beginner does not immediately absorb what he is hearing. *Bravo Alfa Romeo Bravo India Charlie* . . . and so on is meaningless at first, but if this sequence is translated into the letters B–A–R–B–I–C, as it is received, the word begins to have a shape and becomes *Barbican* without difficulty. This process, too, becomes second nature after a while but, initially, it is necessary to make a positive effort to memorize the letter M when hearing the word Mike.

Phonetic figures

With letters there is one phonetic alphabet and it is used worldwide. With figures, on the other hand, there are two versions: the official maritime system and the official aeronautical one. In practice the former is very rarely used.

The words normally used follow the International Civil Aviation code:

Number	Phonetic pronunciation
0	ZERO
1	WUN
2	TOO
3	TREE
4	FOW-er
5	FIFE
6	SIX
7	SEV-en
8	AIT
9	NIN-er
Decimal	DAY-SEE-MAL
Thousand	TOU-SAND

The correct way of giving channel numbers has already been referred to in Chapter 1 and all figures should be transmitted in the recognized manner.

For example:

10	One zero
75	Seven five
100	One zero zero
583	Five eight three
5000	Five thousand
11 000	One one thousand
38 154	Three eight one five four
118.4	One one eight decimal four

The civil aviation authorities recommend the pronunciation given here 'when the language normally used by the station is English'.

It is when the English language is strange that the other (official) phonetic numerals should be used:

Numeral	Code word	Spoken as
0	Nadazero	NAH-DAH-ZAY-ROH
1	Unaone	OO-NAH-WUN
2	Bissotwo	BEES-SOH-TOO
3	Terrathree	TAY-RAH-TREE
4	Kartefour	KAR-TAY-FOWER
5	Pantafive	PAN-TAH-FIVE
6	Soxisix	SOK-SEE-SIX
7	Setteseven	SAY-TAH-SEVEN
8	Oktoeight	OK-TOH-AIT
9	Novenine	NO-VAY-NINER

At first glance those code words might appear somewhat farcical, but they were devised to suit those unable to pronounce certain sounds. The French, for example, find it extremely difficult to say a word like 'moths'; the *ths* combination is almost impossible for them. Because of difficulties with the *th* sound, the number 3 in both tables is pronounced *tree*.

The prefixes introduced before the English words have a logical derivation. *Nada*, for example, means 'nothing' in Spanish. *Una-* and *bi-* have a Latin origin, *penta-* and *octa-* a Greek one. *Karta-* probably derives from the sound of the French *quatre*, and so on.

Punctuation should be used – comma or stop – only when their omission would otherwise cause confusion, but note that a decimal point is spoken as the word 'decimal' not as 'stop' or as 'point'.

Radio log

The regulations used to require that a radiotelephone log was kept and it had to include numerous details as well as the times when the operator was on or off watch. Nowadays the requirement is more realistic, as far as small craft are concerned, and states that 'Each ship fitted with a radiotelephone installation should, where practicable, also carry a radio log.' For normal passage-making R/T entries might be minimal, but there are obvious advantages in having a means of entering the substance of messages in the log in case any distress or urgency traffic is monitored. Entries should take a standard form: time of receipt, identity of calling station, identity of station called, substance of the message and a date time-group identification if one was included.

3 Standard Maritime Navigational Vocabulary

The purpose of having a standard procedure is not to restrict the operator's vocabulary but to ensure the use of the same phrases for the same circumstances. Standard procedure produces a common pattern and the same procedures are now being taught throughout the world. It is because English phrases obviously come so easily to an English speaker that it is doubly important that he should understand the need to use words or phrases in an expected order.

In an article in the International Lifeboat magazine in the late 1970s, the head of the Swedish lifeboat service wrote:

> "Concerning the language barriers in the Baltic where we regularly deal with Finnish, Russian, Polish, German, Dutch and Swedish ships, I think the first contact between operators should always be in English."

And he went on to explain that, once a common language could be used, with common phrases, communication problems were largely solved.

It is probably a mixture of our history of maritime influence and the more recent power of the dollar that has given English its present position. Even French, which for centuries was the diplomatic language of the Western world, is bowing to the use of English.

We who have English as our native tongue do not realize how difficult it can be for others. In face-to-face conversation the 'foreigner' can point or use his hands to make himself understood. With a radiotelephone, he will often be under some stress and thus it is doubly important to use words that will be instantly recognizable to others.

To make my point with a ridiculous example, think for a moment of the word 'foul'.

A man might say 'My propeller is foul', meaning that his ability to manoeuvre was severely restricted. He might say 'My bottom is foul', implying merely that his normal cruising speed was restricted. His report that 'The air in the engine room is foul' could indicate that it was unhealthy or even dangerous to be there. He might refer to 'a foul tide', suggesting a slower progress than might have been expected. He might in conversation speak of 'running foul of the law', with all that that involves. In anger or dismay he might comment 'What a foul thing to say'. He might even be talking about a chicken!

None of these expressions would be likely to confuse another native English speaker, but they could nearly all confuse the foreigner. Furthermore, if the English speaker allowed himself to use slang expressions like 'It was a right old foul up' he could hardly expect to be understood.

Obviously, other examples could be quoted from other languages, but the story may show the importance of understanding what is meant by the use of certain standardized common words and phrases.

In 1976 a committee formed by the then Inter-Governmental Consultative Organization finally agreed the wording of a booklet entitled *Standard Marine Navigation Vocabulary*. It is of interest to remind ourselves that it was first asked for by the Germans and the first draft was prepared by the Dutch! (Merchant Shipping Notice – M.767.) Much of the wording concerns commercial ships only, but there is a great deal that is directly applicable to small craft.

Standard verbs

Where possible sentences should be introduced by one of the following verb forms:

The imperative is always to be used when mandatory orders are given.

YOU MUST	DO NOT	MUST I?
Indicative	Negative	Interrogative
I REQUIRE	I DO NOT REQUIRE	DO I REQUIRE?

I AM	I AM NOT	AM I?
I CAN	I CANNOT	CAN I? CAN YOU?
ADVISE	ADVISE NOT (used when recommendations are given)	

In difficult conditions the interrogative may be preceded by the word 'question': *Question. Can I* . . .

Responses

When the answer to a question is in the affirmative, say YES, followed by the substance of the question if thought necessary.

When the answer to a question is in the negative say NO, followed by the substance of the question.

When information is not immediately available say STAND BY. The station receiving that message would then normally wait on the frequency being used until the station that had answered with STAND BY transmitted again.

When the information cannot be obtained say NO INFORMATION.

Where a message has not been properly heard say SAY AGAIN. (Note that on R/T the word 'repeat' usually has a different meaning.)

Where a message is not understood say MESSAGE NOT UNDERSTOOD.

Urgent messages

(*See* Chapter 6 for examples of the use of the prefixes MAYDAY, PAN-PAN and SÉCURITÉ for distress, urgency and safety traffic. ATTENTION, repeated if necessary, may be used at the beginning of an urgent message.

Miscellaneous phrases

If conditions are poor it may be necessary to ask if your messages are being received. Say: HOW DO YOU READ ME?

I READ YOU
- BAD (1)
- POOR (2)
- FAIR (3) (with signal) strength
- GOOD (4)
- EXCELLENT (5)

- 1 (barely perceptible)
- 2 (weak)
- 3 (fairly good)
- 4 (good)
- 5 (very good)

STAND BY ON CHANNEL . . .

CHANGE TO CHANNEL . . .

I CANNOT READ YOU – PASS YOUR MESSAGE THROUGH . . .

I AM READY (NOT READY) TO RECEIVE YOUR MESSAGE

I DO NOT HAVE CHANNEL . . . PLEASE USE CHANNEL . . .

Repetition

If any parts of a message are considered important enough to need safeguarding use the word 'repeat' in the text. Example:

I HAVE PICKED UP FIVE REPEAT FIVE SURVIVORS

Position

Latitude and longitude can always be used expressed in degrees, minutes and decimals of a minute. When a position is related to a mark ensure that the mark is a well-defined charted object. Bearing shall be in 360 degree notation (true unless stated otherwise) and shall be that of the position *from* the mark. Example:

MY POSITION ONE DECIMAL FIVE MILES ONE NINE ZERO DEGREES FROM START POINT.

Courses

Always in 360 degree notation.

Bearings

Always in 360 degree notation (true unless stated otherwise) *from* the mark or *from* the vessel, except when bearings are relative. Example:

PILOT BOAT IS BEARING ONE SEVEN ZERO FROM YOU.

However, when reporting a position always give the bearing *from* a mark.

Relative bearings
Can be expressed in degrees relative to the ship. Example:
CASUALTY IS ZERO THREE ZERO DEGREES ON YOUR PORT BOW.

Distances
Preferably in nautical miles and tenths of a mile (cables). Kilometres or metres may be used, but the unit must always be stated.

Speed
In knots. Without further qualification this means speed through the water: otherwise give ground speed (speed over the ground).

Time
Use the 24-hour clock, indicating whether GMT, zone time or local time is being employed.

Glossary
The *Standard Marine Navigation Vocabulary* also contains a comprehensive glossary but, as with much of the rest of the booklet, it is aimed at the large vessel. However, some phrases are also of interest for small craft:
Dragging anchor: involuntary movement over the sea bottom.
Dredging anchor: vessel moving under control with anchor moving along the sea bottom.
Established: brought into service/placed in position.
Hampered: vessel restricted in her ability to manoeuvre by the nature of her work.
Height: highest point of the vessel (sometimes called 'air draught').
Inoperative: not functioning.
Track: recommended route to be followed when proceeding between predetermined points.
Vessel inward: from sea to harbour or dock.
Vessel leaving: in the process of leaving a berth or anchorage.

Vessel outward: proceeding from harbour or anchorage seaward.
Way Point: mark or place at which certain vessels are required to report (sometimes called the reporting point or calling-in point).

It is not suggested that all of this brief glossary is of importance to the small craft communicator – and there is a great deal more in the *Vocabulary* itself for those who are interested – but even a knowledge of the existence of some phrases can be useful. It is the foreigner who needs them most and the sometimes insular British must not forget that *they* are foreigners when overseas.

Today the same phrases are being used on R/T in Stavanger, Scheveningen or Sheerness.

Procedural words

It already must be obvious that radiotelephony has a number of special words associated with it. At their worst they amount to jargon and are sometimes misleading; at their best they are useful abbreviations of what would otherwise be exasperating repetitions.

HM Coastguard, the Royal National Lifeboat Institution, the Services, the Civil Aviation Authority and others all have a teaching programme and even though their intentions are similar it is not surprising that some of the procedures in use are different. Thus the procedural words – pro words as they are called – that follow are not 'official' in any sense. They are merely the words likely to be heard on maritime frequencies.

ACKNOWLEDGE: See Received.
ALL AFTER: Used after the pro word SAY AGAIN to request a repetition of all that followed a certain phrase.
ALL BEFORE: Used after the pro word SAY AGAIN to request a repetition of all that preceded a certain phrase.
CONFIRM: MY VERSION IS . . . IS THAT CORRECT?
COPIED: Used by a third station that has monitored an exchange between two other stations and has been asked if she has done so. During a search, for example, A and B might have been exchanging messages and, when the exchange was complete, A (the controlling station) might transmit: C – THIS IS A – DID YOU COPY; the reply being: THIS IS C – COPIED. It is a very useful pro word and translates as: *Yes, do not bother to repeat all that; I was listening.*

CORRECT: The message you have just transmitted was correct.

CORRECTION: Spoken during a transmission it means an error has been made in what the operator has just said. The speaker then goes back to a word or phrase that was correct and continues with the message. Example:
I SHALL BE WITH YOU IN ABOUT FIVE MINUTES – CORRECTION – IN ABOUT ONE FIVE MINUTES.

DATE-TIME GROUP: See Time-group.

IN FIGURES: The following numeral or group of numerals is to be written in figures.

IN LETTERS: The following numeral or group of numerals is to be written in letters as spoken.

I SAY AGAIN: I shall repeat that transmission or a part of it.

I SPELL: I shall spell out the next word or group of words phonetically.

OUT: The end of working to you, the station addressed. It does not necessarily mean the end of my transmission. Nor does it mean that I am ceasing working or ceasing to keep watch. Ideally, at the end of an exchange of messages both stations would add *Out*, to the end of their last reply. In practice, and in good conditions, only one station need say *Out*; the other would not need to reply.

OVER: The invitation to reply. With simplex and semi-duplex equipment it is sometimes necessary to use *Over* at the end of every message during an exchange up to the point when the work is complete, then the final sentence ends with *Out*.

RECEIVED: When a message is received and acknowledgement of its receipt only is required say *Received*. HM Coastguard and the RNLI frequently use the Services jargon 'Roger' (a relic of a now discontinued phonetic alphabet), but the word is not used internationally and its use is to be discouraged.

If the calling station wishes to have a transmission recorded it can end a transmission with:

ACKNOWLEDGE. This is the prerogative of the originator of a message. Example: A – THIS IS B – (message) – ACKNOWLEDGE – TIME NOW ONE FOUR ONE FIVE – OVER. To which the reply might be: B – THIS IS A – YOUR ONE FOUR ONE FIVE ACKNOWLEDGED – OUT.

That use of the pro word 'Acknowledge' reflects the need for important messages to be given a 'time-group' (see below) identity for the log. The time of dispatch thus identifies the message and that message can then be referred to by its time group. Example: . . . MY ONE FOUR FIVE NOW CANCELLED.

Note that *Acknowledged*, in this context, means received and understood.

RADIO CHECK: Another version of HOW DO YOU READ? used by HMCG and by coast radio stations. It is the simplest and best way to get confirmation that your radio is transmitting. If all is well the exchange might be:

Question: THAMES COASTGUARD – THIS IS BARBICAN – RADIO CHECK – OVER.

Reply: THIS IS THAMES COASTGUARD – YOU ARE LOUD AND CLEAR – OVER.

Acknowledgement: THIS IS BARBICAN – LOUD AND CLEAR ALSO – THANK YOU – OUT.

To which Thames Coastguard would not normally bother to reply.

SAY AGAIN: Repeat your last message – or the part of it indicated – I did not get it all. Example: SAY AGAIN THE ADDRESS.

STATION: All ships and shore establishments are called 'stations'. The station initiating the call is the 'calling station' and the other the 'station called'. The word is also used if a vessel receives a call but is uncertain of the name of the originator of the call. Example: STATION CALLING BARBICAN – THIS IS BARBICAN – SAY AGAIN – OVER.

THIS IS: Expression immediately preceding the name and/or call sign of a station.

TIME-GROUP: Four-figure identification for a message giving the time of that message, with an indication whether GMT, zone time, or local time is being used. Time is normally given as GMT unless stated otherwise.

DATE-TIME-GROUP: As for Time-group, but a six-figure identification to include the day of the month before the time. Example: 1125 on the third of the month would be: DATE-TIME-GROUP – ZERO THREE ONE ONE TWO FIVE.

WAIT: If a station called is unable to receive a message it can reply: WAIT ONE MINUTE (sometimes further abbreviated to WAIT ONE). Alternatively, if unable to accept any traffic it would say: *Wait – Out*. In either case the calling station should not call again until she herself was called.

4 Summary of recommended procedure and technique

There are no laws concerning the way a message should be transmitted. The only reason for using a particular method is that departures from it can cause confusion.

VHF Technique

Preparation
Before transmitting, think about the subjects that have to be communicated and, if needs be, prepare written notes to avoid unnecessary breaks or pauses in the traffic.

Repetition
Repetition of words and phrases should be avoided unless specifically requested by the receiving station.

Power Reduction
Use low power (1 W) whenever possible.

Communications with shore stations
Instructions on communication matters from shore stations should be

obeyed. If a change of channel is requested, the receipt of the request must be acknowledged before changing.

Communications with other ships

During intership communications the ship called should indicate the working channel she proposes to use, and the calling ship must acknowledge receipt of the request before changing channel. However, both ships should listen before transmitting on the chosen working channel. If communication is not established on the working channel, change back to the calling channel and then repeat the procedure with another working channel. If communication is unsatisfactory on the working channel chosen, try another, but always await acknowledgement before changing.

Distress

Distress calls have absolute priority over all other communications.

On receipt of a distress message, record it in the radio log and then wait, maintaining a listening watch on Ch. 16. Normally HM Coastguard, or another shore station, will acknowledge distress calls but if there is no acknowledgement and the vessel in distress is in the immediate vicinity, acknowledge receipt. (See also Chapter 6 for Mayday Relay procedure.)

Calling

Whenever possible, a working frequency should be used. Otherwise Ch. 16 may be used if not occupied by distress traffic. When there is difficulty in establishing contact with a ship or with a shore station, allow an adequate interval between calls. Do not occupy a channel unnecessarily with repeated calls.

Watchkeeping

Ships fitted with VHF only should as far as possible maintain a listening watch on Ch. 16. In practice this recommendation means that it is best if a loudspeaker is installed close to the helmsman's normal position. This allows the listening watch to be kept without disturbing those off watch. In certain ports ships are required to keep a listening watch.

VHF Procedure

Say the name of the ship or shore station once (twice only if considered necessary because of poor conditions), followed by the phrase *This is* followed by the ship's name twice, followed, when appropriate, by an indication of the channel in use. Example:

HARWICH HARBOUR RADIO – THIS IS BARBICAN, BARBICAN ON CHANNEL ONE FOUR.

It is appropriate to indicate the channel in use if the station called is likely to be listening on more than one frequency. (See Chapter 5.)

Calling an unknown ship

When calling a ship whose name is not known her position may be used but the call should be addressed to ALL SHIPS:

HELLO ALL SHIPS – THIS IS BARBICAN, BARBICAN – SHIP APPROACHING NUMBER TWO BUXEY BUOY – MY POSITION JUST SOUTH OF RIDGE BUOY.

Caution is necessary when calling unidentified ships, particularly when there is more than one vessel in the vicinity. Sometimes a doubt regarding which ship was being called might be resolved by the use of, say, a white signalling light as a positive identification. (See Chapter 5 for further information regarding All Ships calls.)

Merchant Shipping Notice M.845 suggests that bridge-to-bridge use of VHF for collision avoidance can be dangerous: it can lead to manoeuvres that are contrary to the Collision Regulations and therefore can cause the accident it was intended to avoid.

Acknowledgement

Where a message is received and only acknowledgement of receipt is necessary say RECEIVED. Where an acknowledgement that the correct message has been received say RECEIVED, UNDERSTOOD and repeat the substance of the message if considered necessary:

RECEIVED – UNDERSTOOD – BERTH WILL NOT BE READY UNTIL ONE SEVEN THREE ZERO.

During an exchange of messages a ship invites a reply by saying OVER.

If a message is not properly received, ask for it to be repeated by saying SAY AGAIN.

If a message is not understood say NOT UNDERSTOOD.

If it is necessary to change to another channel say CHANGE TO CHANNEL . . . and then await acknowledgement before changing the channel.

The end of a communication is indicated by the word OUT.

5 Calling procedures

As was explained briefly in Chapter 1, although there are at least 55 channels available with modern equipment, the only call that will ever be received is the call made on the channel or channels being watched (monitored). To call on any other frequency is rather like trying to semaphore to an astronaut: your procedure may be impeccable but the message will never get through.

A call consists of:
- the name or the call sign of the station called, once;
- the words THIS IS;
- the name or call sign of the calling station, twice.

When conditions are bad the name of the station called may be repeated, but never more than three times. In good conditions once is enough because an operator reacts immediately to his own name.

When conditions are good or if the station called is likely to recognize the name of the calling station, there is no need for this to be repeated more than once. But it may be repeated up to three times when necessary.

The beginner tends to gabble names repetitively as if THISISSARAH-JANESARAHJANESARAHJANE was any easier to receive than THIS IS SARAH JANE.

Once contact has been established it should rarely be necessary to transmit any identification more than once. In good conditions abbreviated procedure may sometimes be used but any transmission without identification is forbidden. The call sign may be used – although ships normally use their names except when calling a CRS.

If a station does not reply to a call it may be repeated after three minutes, but repeated calls to a station which may not be in a position to reply, or which may not even be listening, is one of the most irritating and pointless abuses of R/T procedure.

Intership

A basic intership call and transfer to a working channel might be:

BARBICAN — THIS IS FANFARE, FANFARE — OVER.

To which the reply might be:

FANFARE — THIS IS BARBICAN — CHANNEL SEVEN TWO — OVER.

Here, if conditions are good, Fanfare need not use Barbican's name in her reply. Thus she would transmit merely:

THIS IS FANFARE — SEVEN TWO — OVER.

and then both stations would switch to Ch. 72 and Barbican would call — acknowledging Fanfare's initial call, but this time on the agreed working channel.

FANFARE — THIS IS BARBICAN — OVER.

THIS IS FANFARE — GOOD EVENING BERNARD — I EXPECT TO BE . . . and so on.

Note that while on the distress, safety and calling channel there is the absolute minimum of words used. Also, when the working channel has been reached, although Christian or other forenames can be used in the text of a message, the ship's name (or call sign) is still used to identify every transmission.

Unless there has been a prior arrangement, an intership call would normally be made on the distress, safety and calling channel: Ch. 16.

It is the station called that suggests the working channel and she then waits for an acknowledgement before changing. In all official publications Ch. 06 and 08 are listed as the first and second choice for intership working (and Ch. 06 is a 'safe' choice to recommend because all ships have it) but nowadays, with the greatly increased use of multi-channel equipment, it is reasonable to propose the use of a less popular channel (as in the above example), because it is less likely to be already in use. Ch. 06, for example, is used in the Thames Estuary as the Sunk Pilot working channel and Ch. 08 is used a great deal by angling boats, many of whom keep Ch. 08 open all the time they are fishing.

Figure 2 on page 10 shows that there are no fewer than 12 channels designated for intership working, but most have other services allocated to them as well. (For the special uses of channels 06, 67, 72 and 73, see Chapter 8.)

Port operation

Some smaller ports keep watch on Ch. 16 only, but many ports nowadays keep a watch on their working channel as well as on Ch. 16. The official recommendation is now: 'Always call on a working channel if it is known to be watched.' To call on a working channel not only reduces the traffic on Ch. 16, it saves time for both stations.

The *Admiralty List of Radio Signals*, Vol. 6, and the *Admiralty List of Radio Services for Small Craft* list the port operations frequencies and times of watchkeeping. A typical entry from the latter is:

Harwich Harbour   Ch. 16; 11 **14** 17   H24

This indicates that, in addition to watchkeeping on Ch. 16, Harwich Harbour uses three working channels, but Ch. 14 is the preferred one and it is this channel that ships would use for direct calling.

From the small craft point of view, very few ports require calling-in reports from vessels of less than about 50 tonnes. And bearing in mind that the service is normally for commercial shipping, it is unlikely that small craft will need to call a commercial port's harbour radio at all. For a port like Harwich, in contrast to the one-line entry from the *Radio Services for Small Craft*, the *Radio Signals*, Vol. 6, has two pages devoted to Harwich and Ipswich port information. Pilotage, emergency services, pollution control, the harbour limits, radar services and so on are all detailed.

Nevertheless, although a small vessel will not normally call a harbour radio in a commercial port, it can be of considerable interest and help to listen – particularly in bad visibility. The small craft skipper will learn when a particular ship is about to leave her berth, when she will be turning in the channel, when 'So-and-so' is due to reach 'Something' Point, and what the latest visibility reports are. Altogether a useful picture of what is happening is built up.

Bearing in mind the recommendation that, when calling a station where more than one channel is likely to be watched, the frequency in use should be indicated in the initial call, a call to Harwich Harbour Operations Service might be:

HARWICH HARBOUR RADIO – THIS IS BARBICAN – BARBICAN ON CHANNEL ONE FOUR – OVER.

To which the reply might be merely:

BARBICAN – HARWICH HARBOUR

and Barbican would then pass her message directly on that same channel.

Note that there is already a hint of simplified procedure in this exchange. Assuming conditions were good, Harwich Harbour Radio shortened its acknowledgement to 'Harwich Harbour' and the 'This is' has also been dropped. All that is good practice, *once the operators know their job*, but the shortened procedure must never omit the name of the station transmitting. That is both illegal and liable to confuse.

At all times it must be remembered that a port operation radio is a 'net'; dozens may be listening and many will be basing their actions on what they hear. It is often essential that it should be clear to others, as well as to the two stations exchanging messages, who is calling who.

A call is in three parts:
1. the name of the station that is being called;
2. the pro words *This is*;
3. the name of the station that is calling.

Parts 1 and 2 can be dropped, but never 3.

Channel M

As explained in Chapter 1 the frequency 157.85 MHz is licensed to a number of marinas and thus becomes a type of yachtman's port operation channel. The problem is that it is both a calling and working channel and therefore tends to get congested. Also, because it is also used for yacht race administration, there can be difficulties if one man is trying to organize a safety boat while another is booking a berth in the marina. Nevertheless, when there is adequate discipline, the channel is very useful. (See also Chapter 14.)

For 'port' operation on Ch. M the call is direct to the marina office:

SUFFOLK YACHT HARBOUR – THIS IS BARBICAN, BARBICAN – OVER.

BARBICAN – THIS IS SUFFOLK YACHT HARBOUR.

THIS IS BARBICAN – I AM TEN METRES OVERALL WITH A DRAUGHT OF ONE DECIMAL FOUR METRES – CAN I HAVE A BERTH FOR ONE NIGHT? – OVER.

THIS IS SUFFOLK YACHT HARBOUR – YES, BARBICAN, THAT WILL BE ALL RIGHT – GO TO BERTH BRAVO SEVEN – I WILL KEEP A LOOK OUT FOR YOU – OVER.

THIS IS BARBICAN – THANK YOU – MY ETA ABOUT ONE SIX THREE ZERO LOCAL TIME – OUT.

It is very important that the traffic on Ch. M should be kept to a minimum. It is not an intership channel although, regrettably, it is sometimes used as such. If yachtsmen abuse the idea of having the special channel we will end up with the type of silly free-for-all found on Citizens' Band in the United States.

Now that Citizens' Band is legal in the UK it is possible that it could be of considerable help for yacht race management. However, it is too early to say precisely how CB will develop in the UK. (See Appendix C for details.) CB will never replace the marine bands for small craft, but for some specialized purposes it may make an interesting adjunct.

Public correspondence

As we saw in Chapter 1, the expression 'public correspondence' is used for all the ship-to-shore traffic for which there is a charge. Apart from fees for the Ship Licence and for the use of private channels, there is no charge for either intership or for port operation traffic, but all traffic that passes to third parties – traffic which is routed from the radio frequencies into the land-line telephone service – goes via coast radio stations.

The system is international and, in general, it works very well, but one of the problems is that different countries have different types of equipment and different methods of calling. Not unnaturally each administration tends to favour its own method. In the UK, for example, although detailed instruction on how to call a coast radio station (CRS) is given in the *Handbook for Radio Operators* and in the *Notice to Ship Wireless Stations*, neither publication mentions that the procedures may be completely different on the other side of the Channel and North Sea!

Calling a CRS in the UK

With one major exception (1982), all calls to UK coast radio stations are made with an initial call on the distress, safety and calling frequency Ch. 16. The procedure is as for an intership call.
– The name of the coast radio station called once;
– the words *This is*;
– the call sign of the ship, twice.
   Then wait.

A CRS operator will be working with several ships, sometimes on several different frequencies, at the same time. A call may be repeated at about three-minute intervals, but numerous calls do little other than antagonize the operator and exasperate the many other ships monitoring Ch. 16 who will therefore be listening.

Note that all coast radio stations add the word 'radio' after their name. The authorities used to print the name as one word to emphasize this point – NORTHFORELANDRADIO – but they no longer do so, probably because the names became difficult for the foreigner to recognize. Nevertheless, all CRS have the word 'radio' added. If it is omitted from a call the CRS might be confused, with a Coastguard or with a port operation station with a similar name.

> NORTH FORELAND RADIO – THIS IS MIKE ALFA BRAVO ALFA, MIKE ALFA BRAVO ALFA – OVER.

That is a complete call.

North Foreland Radio will answer and give the necessary instructions.

> MIKE ALFA BRAVO ALFA – THIS IS NORTH FORELAND RADIO – CHANGE TO CHANNEL SIX SIX AND STAND BY.

> NORTH FORELAND RADIO – THIS IS MIKE ALFA BRAVO ALFA – CHANNEL SIX SIX AND STANDING BY – OVER.

In practice, and if reception is good, North Foreland might drop the word 'radio' and the words 'change to'.

Similarly stations might shorten their calls slightly so that the exchange was:

MIKE ALFA BRAVO ALFA – NORTH FORELAND – CHANNEL SIX SIX – STAND BY.

THIS IS MIKE ALFA BRAVO ALFA – SIX SIX – STANDING BY.

However, the call should not be briefer than that.

If the CRS was busy, and stations like North Foreland and Niton get very busy indeed at times, the reply might have been:

MIKE ALFA BRAVO ALFA – THIS IS NORTH FORELAND – CHANNEL SIX SIX – YOU ARE TURN NUMBER THREE.

but the end result for the station which initiated the call is the same. Unless there is emergency working (see Chapter 6), or if for some other reason the CRS had to say *Wait – Out*, the ship acknowledges receipt of the instruction to change channel, changes, and then awaits the call on the new channel from the CRS.

One other variation in a call to a CRS might occur if the ship had made the 'wrong' call. In the UK, and in other countries for that matter, many main CRS have 'slave' stations linked to them by land line. For example, Hastings Radio, Thames Radio and Orfordness Radio are all controlled from North Foreland.

If a ship off Harwich had called North Foreland Radio the CRS might have replied:

MIKE ALFA BRAVO ALFA – THIS IS NORTH FORELAND RADIO – WHAT IS YOUR POSITION.

and when told the ship was off, say, the Sunk LtV, North Foreland might switch to her Orfordness Radio aerial off the Suffolk coast and reply:

MIKE ALFA BRAVO ALFA – THIS IS ORFORDNESS RADIO – CHANGE TO CHANNEL SIX TWO AND STAND BY.

If the calling station had been alert she would have checked before making the call and then called the nearest station, be it 'slave' or not.

But often the reception is so good that the CRS can receive on more than one aerial and, if busy in one area, it might switch traffic to work it on another.

About the only other variation in the basic call is that the name of the calling station as well as the call sign is often included in the initial call, but there is no advantage in so doing. The call sign gives the CRS operator all he needs at that stage.

Once contact is made on the working frequency, the CRS will need to know the ship's name as well as what is called the Accounting Authority. For yachts in Britain that would normally be British Telecom itself. For British Telecom the Accounting code is GB 14 and it would be normal for a UK yachtsman to use that as his means of routing the accounts. However, only 18 months after introducing GB 14 as the small craft code British Telecom realized that with a large number of yachts making only a very few calls in any one year, it was costing more than the traffic was worth to collect the dues.

Therefore, from June 1982, a new method of charging was introduced. *It applies only to UK yachts making calls through UK coast radio stations and to UK, Channel Islands or Isle of Man numbers.* Instead of using GB 14 (although that is still correct for other traffic), the new scheme uses YTD as the Accounting Authority Indicator Code. YTD is then followed by the number to which the call is to be charged and the cost of the call then appears (itemized) on the subscriber's normal telephone account without British Telecom having to make out a separate account. There is no additional charge for the facility.

After this initial exchange with the CRS the ship will be standing by on the working channel she had been told to use and, after a short interval, the CRS will call:

MIKE ALFA BRAVO ALFA – THIS IS NORTH FORELAND RADIO.

and the ship replies:

NORTH FORELAND RADIO – THIS IS MIKE ALFA BRAVO ALFA – BARBICAN – I SPELL – BRAVO ALFA ROMEO BRAVO INDIA CHARLIE ALFA NOVEMBER – BARBICAN – ONE LINK-CALL PLEASE – OVER.

At this stage it is best to await an acknowledgement that the CRS can handle a call because, until this moment, the CRS did not know what service it was required to give.

BARBICAN – THIS IS NORTH FORELAND – WHAT NUMBER DO YOU WANT?

and the reply:

NORTH FORELAND – THIS IS BARBICAN – ONE LINK-CALL PLEASE TO ZERO SEVEN ZERO SEVEN THREE TWO SEVEN FIVE SIX FOUR – THE ACCOUNTING AUTHORITY CODE IS YANKEE TANGO DELTA ZERO SIX TWO ONE SEVEN EIGHT FOUR ONE ZERO THREE – OVER.

BARBICAN – THIS IS NORTH FORELAND – STAND BY.

Note that this example uses the YTD Accounting Authority Indicator Code because the call is from a UK yacht to a UK number. In other circumstances, or when overseas, the UK-registered yacht would normally use GB 14 as her indicator code.

Note also that the CRS operator, because he is a professional and used to working on duplex equipment where there is no press-to-speak switch probably does not use *over* when answering the initial call and certainly will not say *over* after *stand by* because this phrase means what it says; it is not an invitation to reply. It would now be the CRS that would call next, after the link had been made.

However, even if the CRS operator – who can transmit and receive simultaneously – does not often use *over* it is wise for the yachtsman using simplex equipment always to go by the book and to include it. It soon becomes second nature and, once adopted as a habit, it avoids ambiguity when speaking to non-professionals.

The UK exception

Everything in the above section applies to *nearly* all UK coast radio stations, and to many foreign stations as well, but the idea of calling a CRS on Ch. 16 is becoming outdated.

The principle of calling on the distress, safety and calling channel, the frequency on which all ships are expected to keep watch, works only if there is not too much traffic on it. Thus there has been a move away from calling a CRS on Ch. 16. Niton Radio, on the Isle of Wight, was the first United Kingdom CRS to install new equipment to allow calls to be accepted direct on the working channels.

From the beginning of 1981 calls to Niton Radio have been made by switching to any of Niton Radio's several working channels; waiting until one is free and then calling direct. The fact that a channel is busy is indicated by an engaged signal or by speech.

Other United Kingdom CRS will follow the practice as the new specialized equipment becomes available and by the end of 1983 nearly half of the UK CRS will have changed.

Calling coast radio stations outside UK waters

It must be admitted that several northwest European countries were ahead of the UK regarding the type of equipment installed. The reasons were sometimes geographical – a short coastline for example – but the fact remains that in the Netherlands, calls to Scheveningen Radio (the call sign for the entire Dutch coast) have been made on an appropriate frequency for several years. The choice of working channels depends on the ship's position but, having looked up the right channel, the ship calls once, on the appropriate working channel, and then waits. The fact that the call has been accepted, and stored in a computer, is notified by a jingle, which repeats itself at intervals until, if the call has not been answered after about 3 minutes, the ship may call again.

What is clever about the Dutch method is that the call is always the same – always Scheveningen Radio – but the traffic is then switched automatically to the desk of whichever operator is next free. This method has obvious advantages for both sides and it suits a small country with a short coastline and just one central coast radio station. The Dutch still monitor Ch. 16 and will still accept calls on that frequency, but they prefer ships to call direct.

In neighbouring Belgium there is also only one call sign on the coast – Oostende Radio – although there are three different frequencies used, depending on the ship's position. The Belgians also accept calls on Ch. 16, in the same manner as they are received in the UK. In France, however, there is yet another variation. Calls are made direct on the appropriate working frequency merely by holding the press-to-speak key down for three or four seconds (to transmit a carrier wave) and then listening. A ringing tone will be heard when the call is accepted and the French CRS will then transmit, asking for the name of the ship that is calling.

The big difference in France is that there is no calling on Ch. 16 – in fact the CRS that handle public correspondence do not monitor it. On West Germany's North Sea coast calls to Norddeich Radio may be made on Ch. 16 *or* on the Norddeich working channels 25, 28 and 61.

From all this it is clear that western Europe has many different methods. For the West German Elbe–Weser stations, calls should be on one of three working channels. Sweden accepts calls on working channels only. Denmark and Norway have recently changed to accepting calls on working channels only.

There are no universal rules. See Appendix B for further details.

Duration of calls and time limits

When a link-call is complete it is best to wait on the working frequency that has just been used for the call because, normally, the CRS operator will call back giving the duration of the call just made. There is a minimum three-minute charge for all calls from CRS. Also, when the channels are congested, the coast radio station may limit the duration of all calls to six minutes.

CRS calls to ships

To make a call to a ship via the land-line telephone service the caller (in the UK) can dial 100 and ask for 'Ship's Telephone Service', and he will be put through to Portishead Radio – the long-range British Telecom station. The operator will need to know the ship's name, the call sign if known and, if there is one, the selective call number (see below). He will also need an approximation of the ship's position so that he can transfer the caller to the nearest CRS.

There are three ways in which ships are contacted by the CRS.

Traffic lists: Every two hours throughout most of the 24, and for seven days a week, all CRS broadcast traffic lists – lists of the names of the ships for which it holds traffic. Ships are expected to listen to these lists, which are broadcast on the station's principal working frequency and, in most countries, after a brief preliminary announcement on Ch. 16. On hearing that there is traffic for her, the ship then calls the CRS in the normal way, as soon as she is ready to receive the traffic.

Direct calls: In addition to the regular traffic lists, a CRS will often call a ship direct if there is traffic waiting, but obviously the CRS is unlikely to do that unless it was known that the ship was within range and was likely to be keeping watch. If this applies, the call, on Ch. 16, would be:

BARBICAN– THIS IS NORTH FORELAND RADIO – TRAFFIC FOR YOU.

Traffic lists are preceded by the pro words *All ships, All ships, All ships*. During busy periods, the CRS may make additional *All ships* calls, followed by a short list of names of ships for which they are holding traffic. It is obvious that if the ship is likely to be keeping watch it is in the interests of everyone to clear the traffic straight away rather than hold it for what might be a much busier period immediately after a routine traffic list.

Selective calling: The third way in which a CRS can call a ship is with special equipment which can be used only if the R/T set on the ship has been fitted with a decoder. Selective calling is usually an optional extra but can be standard equipment on the more sophisticated sets. However, selective calling has such tremendous advantages for small craft that it is worth describing the system in some detail.

The principal advantage that 'selcall' has is that it becomes a private paging system, when necessary, and one for which there is no charge. Whether a vessel is under way, in harbour or in the middle of a complicated manoeuvre it must be obvious that there will be occasions when a call from a CRS might not be heard or, if it had been heard, it might not have been possible to answer. With selcall, although the call is transmitted on Ch. 16 – the same channel as would be used for a direct call – the selcall code goes out automatically (in milliseconds) and it triggers a light and an alarm buzzer on board.

First the buzzer draws attention to the fact that the vessel is being called by a CRS but, more important for the small craft skipper, although the buzzer goes off after a few seconds, the light remains. So even if the vessel was unmanned when the call was received – the skipper might be ashore – the fact that there had been a call is recorded. All the vessel has to do is to cancel the light on the selcall equipment and then call the CRS in the normal manner and say 'You had traffic for me.'

Of course, the VHF set has to be on, and tuned to Ch. 16, but a further advantage is that on most sets the loudspeakers can be muted; the set remains on but the speaker is not 'bleating away' all night. If for any reason a call was expected, the VHF can be left on while the crew are asleep in a quiet anchorage, relaxed in the knowledge that if the call should come, the equipment will wake them.

A CRS will have a record of all ships fitted with selective calling but if, when booking a link-call, the caller knew the ship's selective call number – five digits – it would save time for him to give it. All the CRS operator then has to do is to tap in the selcall number into his encoder and the call is transmitted twice, automatically, with a 3-millisecond interval between the two calls. If the ship did not call back after a selcall transmission, the fact that the CRS held traffic for her would still be broadcast on the next two traffic lists.

There is yet another advantage of selcall: in an emergency the CRS can switch the shore-based encoder to a different mode and trigger the alarms for all ships within range. Thus for an urgent navigational warning, or for distress, all ships with the equipment can be alerted.

There is only the one alarm buzzer on the ship's set, but two different warning lights: one for the call to the ship alone and the other for an 'all ships' call. The latter is termed a CQ call – CQ meaning 'All stations' in the International Q Code used by professional operators. On many sets the twin lights are labelled 'Call' and 'CQ'.

Selective call code numbers are allocated by the licensing authority – the Home Office in the United Kingdom – and there is no charge other than the licence fee itself. A vessel with selcall has therefore two unique identifications: her international call sign and her selective calling code. The former is for use in any circumstances but the selcall code is for use solely by CRS and then only when it is the CRS that wants to contact the ship: it is not needed for traffic in the opposite direction.

Distress and urgency calls

Having dealt in this chapter with intership, port operation, channel M and all the types of public correspondence calls, the only major omission is distress traffic. However, there is one overriding difference between everything that has been said about calling so far, and what follows.

All distress and urgency traffic takes place on Ch. 16, the distress, safety and calling channel: both the call and the messages. This is a fundamental difference, and it is so important that distress procedures should be unambiguous that they are given a chapter of their own.

6 Distress and Urgency procedure

- The use of the distress signal is absolutely forbidden except in the case of distress.
- The distress signal is provided for use in cases of imminent danger when immediate aid is necessary. Its use for less urgent purposes might result in insufficient attention being paid to calls made from ships which really require immediate assistance.
- Where the sending of the distress signal is not fully justified, use should be made of the urgency signal which has priority over all other communications except distress.

Observant readers will have noticed that, almost without exception, whenever Ch. 16 has been mentioned the fact that it is also the distress and safety channel as well as the calling channel has been added. The point is vital to an understanding of maritime VHF radiotelephony.

A large vessel will have either a professional operator on duty in the radio room or, on ships that carry only one man, there is special equipment tuned to the MF distress frequency which triggers an auto-alarm on the bridge and in the wireless officer's cabin.

For ships not carrying professional wireless operators, there is also an auto-alarm system on the MF distress channel, and that, too, triggers buzzers if a special two-tone distress alarm signal is received.

On VHF, on the other hand, there is no auto-alarm system and therefore small vessels that rely solely on VHF equipment have an even greater need to observe discipline. If Ch. 16 is occupied while someone rambles on making a muddled call it is not available for distress traffic.

Carrier waves

Inside the front cover of this book is a list of 'Ten Golden Rules'. As far as distress working is concerned the tenth rule is perhaps the most important: 'When the traffic is complete, ensure that your press-to-speak switch is released and that a carrier wave is not being transmitted when the microphone is in its cradle.'

When under stress it is all too easy to throw a telephone handset down on to a settee, or to let it fall behind some books. However, if the press-to-speak switch, which is usually only very lightly spring-loaded, falls against anything, the transmit side of the set will become alive and the receive side will go dead. Not only is it impossible to call a ship and say 'Excuse me, but I think you might be transmitting a carrier wave', but the fact that the ship is transmitting will blot out any other transmissions in the immediate vicinity anyway. It is a 'heads neither of us wins; tails we both lose' situation.

The effect can be so devastating that on occasions the BBC has been asked to broadcast a warning, on normal BBC channels, to ships in a certain area saying: 'Will you please check to ensure you are not transmitting a carrier wave inadvertently.'

A 'carrier wave' is just noise, not speech, but it blots out other traffic as effectively as if the culprit was singing 'What shall we do with the drunken sailor?'

Distress

In previous chapters everything written about procedures has been based on the requirements for VHF radiotelephony. However, although there is a little information about medium frequency R/T in Appendix D, it is worth bearing in mind at this point that 2182 kHz is the international distress, calling and safety frequency for MF R/T, and that most maritime receiving radios can be tuned to receive on that frequency. If a small craft is involved in any way with distress working, it can be of considerable help to keep her receiver tuned to 2182 kHz. That will be the frequency the CRS uses for distress traffic and for its relays to other ships, and the use of MF for reception to back up the VHF traffic might make an important difference.

Apart from that, it is Ch. 16 – 156.8 MHz – that is the VHF distress frequency, and all calls as well as all traffic takes place on that channel.

All books about radio as well as the leaflet sent out with all ship licences emphasize that distress traffic must be transmitted in a particular manner, but few say why. However, once the reason is understood it is very much easier to remember the procedure.

Note carefully: 1. a distress *signal*, 2. a distress *call*, 3. a distress *message* are three different things – connected but different.

As already explained, there is no auto-alarm distress signal for VHF; the **signal** is the word 'MAYDAY'.

The distress **call** is:
– the distress signal MAYDAY, spoken three times;
– the words THIS IS;
– the name (or other identification) of the vessel, spoken three times.

The distress **message** is:
– the distress signal MAYDAY (once);
– the name (or other identification) of the vessel in distress (once);
– the particulars of the position;
– the nature of the distress and the kind of assistance desired;
– any other information that might facilitate the rescue;
– the invitation to reply.

For example (and remembering that there is no auto-alarm) the complete distress call and message might be:

The distress call

– distress signal (three times)	MAYDAY MAYDAY MAYDAY
– the words THIS IS	THIS IS
– the name of the ship (three times)	FANFARE FANFARE FANFARE

The distress message

– distress signal	MAYDAY
– name of ship	FANFARE
– the position	ONE MILE EAST SUNK HEAD TOWER
– nature of distress	HIT UNLIT BUOY MAKING WATER

– other information	UNLIKELY TO BE ABLE TO KEEP PACE WITH LEAKS WILL FIRE DISTRESS ROCKETS AT FIFTEEN MINUTE INTERVALS
– invitation to reply	OVER

The reason why a call and message is spoken in this particular manner will be more easily understood in a moment.

Acknowledgement of distress

If a distress call is heard the initial reaction should be to log the receipt of the call and do nothing. In coastal waters distress calls will be acknowledged by a coast radio station or, if on Ch. 16, by the Coastguard. It is merely confusing if others start to acknowledge, but a distress call must be acknowledged and if, for any reason, no acknowledgement is heard, then the vessel receiving the call should herself acknowledge.

An acknowledgement, too, has an internationally agreed form:
– the distress signal MAYDAY;
– the name of the vessel sending the distress, spoken three times;
– the words THIS IS;
– the name (or other identification) of the vessel acknowledging receipt, spoken three times;
– the word RECEIVED;
– the distress signal MAYDAY.

For example:

MAYDAY – FANFARE FANFARE FANFARE – THIS IS BARBICAN BARBICAN BARBICAN – RECEIVED MAYDAY.

This transmission, more than the previous example, begins to show the logic of the procedure. The word MAYDAY, which comes from the French *m'aider* (help me), is the distress signal. It alerts all who hear it that the message that follows has priority.

The initial call repeats the signal – the word MAYDAY – three times, and the ship's name is also repeated three times.

The signal (the word MAYDAY) is also transmitted as the first word of the distress message, and that is followed by the name of the ship in distress, also once. Thus all distress messages start with the signal; this 'magic' word MAYDAY.

An acknowledgement of the receipt of a distress message is itself a distress message so it, too, starts with MAYDAY (as in my example) and the same applies to any other message during the exchange.

Note, too, that the acknowledgement is just that: it does not end with *Over* because it is not an invitation to reply.

Every station that has acknowledged a distress message should, as soon as possible, follow up the acknowledgement, but the first job is to let the ship in distress know that her call has been heard. The follow-up call could let the ship in distress know the position of the vessel that had acknowledged her distress, the speed that she could make towards the casualty, and the approximate time it will take her to get there. From this it is obvious that there is little point in rushing to acknowledge a distress call unless you are sure that you are likely to be in a position to help.

Normally, distress calls on Ch. 16, will be acknowledged by HM Coastguard and it is they who subsequently control the exchange of messages.

Mayday Relay

Distress messages might have to be relayed for any of three different reasons: 1. when the station in distress cannot herself transmit a message; 2. when the master of a ship or the person responsible at the shore station considers that further help is necessary; 3. when a ship, although not herself in a position to render assistance, has heard a distress message that has not been acknowledged.

Obviously the fact that the ship transmitting is not herself in distress must be made clear – especially these days with the increasing use being made of VHF D/F equipment by the rescue authorities. Thus a Mayday Relay also has a precise form of words:

– the signal	MAYDAY RELAY MAYDAY RELAY MAYDAY RELAY
– the words	THIS IS
– the name or call sign of the vessel making the transmission	BARBICAN BARBICAN BARBICAN

This is followed by the distress message, which also starts with the distress signal MAYDAY:

- the distress signal   MAYDAY
- name of ship in distress   FANFARE
- the position   ONE MILE EAST (and so on)

Normally the receipt of a Mayday Relay would be something that should be logged but no further action would be required. However, if on hearing a Mayday Relay a vessel realized that the vessel in distress was in her vicinity, she might wish to call the station controlling the distress traffic to let that station know her intentions. This is also classed as distress traffic and it too takes a particular form.

Distress traffic

The likelihood of small craft being involved in actual rescue traffic on VHF radiotelephony is statistically small, but if she was, the efficiency of her procedure might well make the difference between life and death. That is not an over-dramatization. During distress working there is often a great deal of traffic, from the casualty herself, other ships in the vicinity which might be converging on the area, the shore station trying to control what is happening, other ships transmitting (contrary to the radio regulations) because they were not aware that there was distress working and, finally, the small craft that happens to be in the immediate vicinity and might be in a position to relay because she could hear the messages from the casualty although the shore could not. She might have been monitoring the distress traffic and felt it right to let the station controlling that traffic know that she was in the immediate vicinity.

All distress traffic is preceded by the distress signal MAYDAY both before the call and before the message:

MAYDAY − THAMES COASTGUARD − THIS IS BARBICAN BARBICAN − MAYDAY FANFARE − MY POSITION IS . . . AND I PROPOSE TO . . . (and so on).

The first use of *Mayday*, at the start of the call, alerts the station to which the message is directed, as well as all other stations listening, that what follows is distress traffic. The second use of the word *Mayday* comes at the start of the message and identifies what follows as a message associated with the *Fanfare* distress traffic. After all, the shore station handling the *Fanfare* traffic might be involved in some other distress traffic at the same time.

Urgency calls

The second priority, which has precedence over everything except distress traffic, is for urgency messages concerning the safety of a ship or the safety of a person. The distinction between distress and urgency is important. Distress involves imminent danger to the vessel and the need for immediate assistance; urgency involves the safety of a vessel or of the persons on board.

The urgency *signal* is three repetitions of the words *Pan-pan* (derived from the French word *panne* meaning breakdown) and urgency messages can be addressed to 'All stations' or to a particular station. An urgency *signal*, *call* and *message* might be:

– urgency *signal* PAN-PAN – PAN-PAN – PAN-PAN
– *call* name of station called (up to three times) HELLO ALL STATIONS – ALL STATIONS – ALL STATIONS
– the words THIS IS THIS IS
– name of calling station (up to three times) FANFARE FANFARE FANFARE
– urgency *message* FIVE MILES SOUTH START POINT – DISMASTED AND ROPE AROUND PROPELLER – NEED TOW
– INVITATION TO REPLY OVER.

As with a distress message, the urgency message has a particular form: 1. position, 2. nature of urgency, 3. assistance required. It ends with the invitation to reply because the object is to alert any nearby vessel that might be able to offer assistance. Alternatively, an urgency call might be picked up by HM Coastguard who could arrange help.

The Royal National Lifeboat Institution is not a towage service and would not normally be called out merely because a yacht was dismasted. However, in practice – and certainly if HM Coastguard felt that the safety of the crew was at risk – the RNLI would be involved if a yacht was dismasted and needed a tow. Distress traffic always takes place on

Ch. 16, but urgency traffic *may* change to a working channel if there are long messages to relay.

Safety calls

The third degree of priority is indicated by the safety signal consisting of the French word *Sécurité* (pronounced SAY-CURE-E-TAY) sent three times. It precedes transmissions concerning important navigational or meteorological warnings and is therefore normally used by coast radio stations and addressed to 'All stations'.

The safety signal is normally transmitted on Ch. 16, but the message should be on a working channel.

Technically, there is a difference between an 'All ships' call, which on HF and MF is for distress and urgency traffic, although it may be used on VHF for safety purposes, and an 'All stations' call. In practice the distinction has become blurred, but all ships should listen to urgency and safety calls for long enough to satisfy themselves that the message does not concern them and they must avoid interfering with any message that follows.

Control of distress traffic

Initially the control of distress communications is the responsibility of the ship that is in distress, but in practice the responsibility is usually delegated to or assumed by a shore station. For coastal traffic on VHF HM Coastguard has the responsibility for keeping a distress watch on Ch. 16 and also for co-ordinating any search and rescue operations. (See Chapter 8.)

If commercial shipping is involved, and if the distress traffic has been on the MF distress frequency 2182 kHz, the coast radio stations would also be involved because they are better equipped to handle any MF relay traffic to ships. It is the British Telecom CRS that has the responsibility for the distress watch on both 2182 kHz and 500 kHz because in any major incident other ships will also be involved.

Nowadays the Services are also likely to be called in, so yet another authority can be involved, but that does not alter the initial coordinating role of HM Coastguard.

In the event of extensive Service participation one or other of the two military Rescue Co-ordinating Centres at Pitreavie, in Scotland, and at Plymouth, in the West Country, will on request assume responsibility. It is they who, on behalf of the Civil Aviation Authority and in accordance with the arrangements laid down by the International Civil Aviation Organization (ICAO), undertake search and rescue operations for aircraft in distress, but the transfer of responsibility from HM Coastguard to the military Rescue Co-ordinating Centres, occurs only for major incidents.

In these circumstances, when specialized search and rescue ships as well as merchant vessels are on the scene, one of the specialized units will assume the duties of On Scene Commander and will direct search operations.

In more normal circumstances, when Service units are engaged in civil SAR operations co-ordinated by HM Coastguard, the Service helicopters or other craft report to the Coastguard Maritime Rescue Co-ordinating Centre or Sub-Centre, although they are still technically under Service operational control.

Imposing silence

On numerous occasions in this book the need for discipline on the air has been emphasized and it must be obvious that during distress working that need is greater than ever. The station controlling distress traffic may impose silence either on all stations of the mobile service in the area or on any station that is interfering with distress traffic. Silence is imposed by the station which is controlling the distress transmitting:

SEELONCE MAYDAY – THIS IS THAMES COASTGUARD – OUT. (*Seelonce* represents the French pronunciation of 'silence'.)

No other station may use that expression, but if another station in the vicinity believes it is essential to do so she may transmit:

SEELONCE DISTRESS – THIS IS BARBICAN – OUT.

When complete silence is no longer considered necessary the station controlling the traffic transmits an 'All stations' call:

– the distress *signal*	MAYDAY
– the *call*	HELLO ALL STATIONS – ALL STATIONS – ALL STATIONS

– the words THIS IS	THIS IS
– the name or other identification of the station sending the message	THAMES COASTGUARD
– the time (date-time-group of the Mayday to which this message refers)	ZERO THREE ONE ONE FIVE ZERO GMT
– the name or call sign of the vessel in distress	FANFARE
– the word PRU-DONCE	PRU-DONCE.

This indicates that restricted working, confined to essential calls, may be made, but great care must be taken to avoid interference with subsequent signals from the casualty, which should now be prefixed with the urgency signal.

Finally, when all the distress traffic has ceased, the station that has been controlling the traffic must let all stations know that normal working may be resumed. She does so by sending:

– the distress *signal*	MAYDAY
– the *call*	HELLO ALL STATIONS – ALL STATIONS – ALL STATIONS
– the words THIS IS	THIS IS
– the name of the station sending the message	THAMES COASTGUARD
– the time (date-time-group of the *Mayday* to which this message refers)	ZERO THREE ONE ONE FIVE ZERO GMT
– the name or call sign of the vessel which was in distress	FANFARE
– the words SEELONCE FEENEE	SEELONCE FEENEE

All this distress procedure may appear somewhat complicated. Those who have English as their first language must remember that the procedure, and its precise wording is the result of international agreement. A Russian talking to a German will still be using expressions like *Silence finis* and as far as possible the pronunciation will be as shown.

The official publications lay great stress on how a casualty should transmit – and, of course, this is extremely important – but the whole complicated structure was planned on the assumption that most of the operators would be professionals and that professionals would be handling the subsequent traffic. Nowadays, in coastal waters at any rate, the vast majority of those listening to any distress communications will be amateurs and therefore it is vital for them to be able to understand what is going on and when, if at all, it would be right for them to become involved in the traffic.

This chapter has been condensed from the *Handbook for Radio Operators*, the appropriate section in *Admiralty Notice to Mariners* and from the HM Coastguard and RNLI training manuals. It needs to be studied carefully because once the reasons for following an internationally agreed procedure are understood, then that procedure is all the easier to commit to memory.

When a man needs to follow distress or urgency procedures he will be under stress and that would not be the right moment to start to 'read up' what he ought to be doing.

The endpaper of this book is designed as a memory aid.

7 Jargon

All trades and all professions tend to breed their own jargon and the effect it can have, both good and bad, is neatly summarized by the dictionary definition of the word: 'Jargon. Unintelligible words, gibberish, or debased language; mode of speech familiar only to a group or profession.'

It can be further subdivided into:
1 useful abbreviation;
2 confusing brevity;
3 jargon that is the very opposite of brevity;
4 the misuse of procedural words.

Useful abbreviation

Examples of useful brevity have appeared in many places already. In fact most of the section concerning the use of pro words comes in that category. It is many times simpler to say 'RADIO CHECK' than to have to say 'WILL YOU LET ME KNOW HOW YOU ARE RECEIVING MY TRANSMISSION?' It is jargon in the sense that the use of the expression is not laid down in any international recommendation (although it is taught, internally, by HM Coastguard), but it is sensible jargon that is reasonably self-explanatory.

Confusing abbreviation

There are countless examples from professionals. Pilots, for example, using VHF R/T for 'bridge-to-bridge' traffic often use verbal shorthand such as 'GREEN TO GREEN'. That would apply if, for any reason, one pilot was suggesting to the pilot of another vessel that they should pass 'starboard side to starboard side' – contrary to the normal Collision Regulation when two vessels meet.

'GREEN TO GREEN' is an expression often heard at a port like Harwich where commercial traffic, which has to keep to a dredged fairway, makes a sharp turn on entering the harbour. Apparently there are conditions when it suits both ships to pass starboard to starboard. The expression is helpful in that it is both a question and an answer. One pilot will say, merely: B – THIS IS A – GREEN TO GREEN, and the acknowledgement would be simply: THIS IS B – GREEN TO GREEN.

The danger in this type of exchange is that others who might be listening do not necessarily understand what is being said. Furthermore, once the expression has become familiar to one group – in this instance the pilots – there is a danger that they use it when speaking to others who do not fully understand the implications.

There are plenty of other examples of confusing brevity and most originate among professionals who wish to avoid what to them is meaningless repetition. A Belgian ferry might pass a visibility report at the 'MPC buoy' (Mid-Pas de Calais). The Townsend ferry *Free Enterprise IV* will be referred to by North Foreland Radio operator as 'FE 4', and so on: useful when the exchange is solely an exchange between fellow professionals, but dangerous in the sense that jargon tends to spread.

Verbosity

Verbosity is so common that it spreads into everyday speech. People who cannot stop themselves saying, 'At this moment in time' when they mean 'now' or 'today' can be a menace, but the addition of superfluous words in R/T is even more habit-forming. 'WHAT IS YOUR POSITION AT THIS TIME? CHANGE TO CHANNEL "X" IF YOU HAVE IT FITTED.' 'B – THIS IS A – DO YOU READ?' or 'B THIS IS A – ARE YOU ON THE AIR?'

Unfortunately there is a tremendous amount of this type of meaningless padding in radiotelephony. Among the older professionals the habit is probably a relic of the use of aerial tuning knobs on MF radio when it was necessary to tune in to a call. However, the use of expressions like 'THIS IS B, DO YOU READ?' on VHF R/T merely converts a three-word sentence into a six-word one. The addition does not mean anything because the phrase is not being used to ask for a 'radio check'. It is a superfluous piece of waffle akin to saying 'ARE YOU THERE?'

The misuse of procedural words

On a TV serial the wrong use of R/T pro words does not really matter – except that people are liable to copy what the 'hero' says – but their misuse among professionals on maritime radio can be dangerous. The amateur, especially when he is learning, desperately wants to appear efficient and professional in his manner and he is therefore likely to pick up and spread the use of bad verbal habits.

One of the more unfortunate confusions concerning otherwise useful jargon is the misuse of the pro word 'copy' or 'copied'. In the Services a pilot might be given an instruction that he would have to write down. Having done so he might say ALL COPIED meaning that he had actually written it all down. HM Coastguard, on the other hand, teaches a completely different use for 'copy' and it can be a great time-saver when organizing a search and rescue operation.

It has been stressed several times that VHF R/T is a 'net', with dozens listening for any one transmission. During the organization of a search one station might be passing information to another and, when the message has been completed, the station called would reply RECEIVED – OUT. However, the calling station might then transmit to yet another station, or more than one other station. For example: BARBICAN – FANFARE – THIS IS THAMES COASTGUARD – DID YOU COPY? and, if they had done so both stations (in the order in which they had been called) would reply: THIS IS BARBICAN – COPIED – OUT. THIS IS FANFARE – COPIED – OUT.

By that use of the pro word 'copy', Thames Coastguard is able to say 'Did you listen to and understand all the message I have just sent, or do I have to repeat it all again?' All that by the one word 'COPY'.

The word is misused when a station called substitutes the word 'COPY' for 'RECEIVED', which is properly the acknowledgement of a message sent to the station. 'COPIED' is an acknowledgement of a message sent to another station but also received (overheard) by the third party, which is now acknowledging.

Finally, one of the most insidious of all pieces of jargon: the use of

'Roger' instead of 'Received'. This is a relic of the days when 'Roger' stood for R in the old phonetic alphabet. It is so well-established in Services R/T that it is still taught there, and by HM Coastguard. However, it is not taught anywhere else and does not appear in any international manuals and is bound to confuse if used in other than internal traffic. The use of 'Roger' was discussed in an Inter-Governmental Maritime Consultative Organization subcommittee early in 1980, when the Standard Maritime Navigation Vocabulary was under review. The United Kingdom suggested that it should be used as a pro word but, not unreasonably, the rest of the world objected. 'Roger', unfortunately, is so ingrained in some Services jargon that it will remain in British use for some time, but as it has no international currency it should be discouraged.

8 HM Coastguard and communications

In Britain the role of HM Coastguard has changed considerably at different times; the most recent change was a reorganization of its resources that began in 1978. Today the full-time professional staff is usually concentrated in Maritime Rescue Co-ordination Centres and Sub-Centres, of which there are 22 in all, and the Auxiliaries man the look-outs and the mobile stations.

There are about 600 full-time Coastguards supported by 9000 Auxiliaries and although during the most recent reorganization the co-ordinating role has changed little, the way in which the co-ordination is done has been altered dramatically.

HM Coastguard's prime task is initiating and co-ordinating all civil marine search and rescue measures for both vessels and persons in need of assistance throughout the United Kingdom Search and Rescue Region. Co-ordination requires efficient and reliable communications and the modern Coastguard is more likely to be working with radiotelephones and Telex machines from a communications centre than he is to be peering through a telescope from a hut on the top of a cliff. The cliff-top look-out, when one is maintained, will be the job of an auxiliary, but he too is more likely to be operating from a radio-equipped four-wheeled drive vehicle than from a hut.

Channel Zero, 67 and 73

As has already been outlined in Chapter 1, in the UK HM Coastguard has the use of Ch. 00 (zero) 156.00 MHz as the primary channel for communication. As second choice, it will switch to Ch. 67 for search and rescue working (in addition to its use for direct communication with small craft as described in Chapter 1), and then to Ch. 73 as third choice.

Ch. 00 is just outside the international maritime band (and its use has to be separately licensed). Ch. 67 and 73 are within the international maritime band and are allocated, internationally, for intership and port operations traffic. Therefore, as with many other aspects of R/T procedure, there has to be an understanding and a degree of give and take. HM Coastguard does not have the power to order radio silence on Ch. 67 or 73, but there are occasions when ships may be requested to change to another working channel because of essential SAR or safety traffic on those two channels.

Ships normally have several other channels from which to choose; HMCG is restricted to those two alone.

During distress working, of course, the traffic is normally on Ch. 16 and has the necessary priorities, but it is during multiple casualties, exercises or other similar occasions that HMCG may need Ch. 67 and 73; one obvious reason being the existence of distress working in a neighbouring region.

HMCG has exclusive control over the use of Ch. 00 because of the need to communicate with so many different agencies: it would not be practical to risk interference. During an actual SAR incident, the Coastguard would normally communicate with the casualty, with any RNLI lifeboat, with any SAR helicopter, and with any other vessel that might be involved, on Ch. 16. But at the same time the Rescue Centre might be in touch with other SAR helicopters, other Coastguard Centres, or vehicles, possibly several Auxiliary Coastguards, with police boats or Trinity House vessels. That traffic would normally take place on Ch. 00. The normal land-line telephone service would be used for communicating with the ambulance service, doctors, local hospitals, shore-based police, the fire services, members of the public and, in any major incident, the press.

During any incident the skill and the bravery of those 'on scene' is as important as ever it was. The ability to communicate, however, has made it possible to concentrate so much more effort onto any one incident than was ever possible in the past. The co-ordination of the

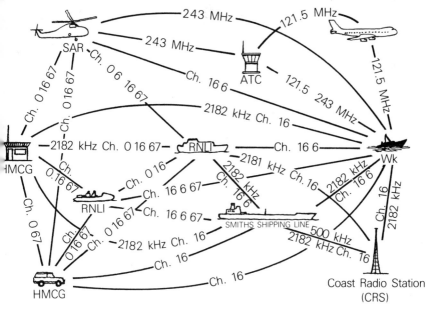

Figure 3 *As far as a small vessel in distress is concerned all radio traffic is likely to be on VHF Ch. 16 but there is a great deal more to the problem than that. This 'cat's-cradle' shows the frequencies normally used during a distress or urgency incident and it explains, if explanation is needed, why efficient co-ordination of the various activities is important.*

work falls on HMCG and the bulk of the communication is by radiotelephone. (See figure 3.)

Precedence indicators

Chapters 1 and 6 explained the use of the distress, urgency and safety precedence indicators – MAYDAY, PAN-PAN and SÉCURITÉ – which impose different degrees of radio silence or discipline on those who hear them. HM Coastguard, on the other hand, has somewhat different traffic priorities and it uses indicators of its own.

Normally the traffic between HMCG stations – fixed or mobile – is on their private frequency, Ch. 00, but it also takes place on maritime frequencies and so it helps to understand a little of Coastguard procedure, even if yachtsmen are unlikely to need to use it.

A station calling Coastguard may use PRIORITY as the prefix for a call that might have used SÉCURITÉ if it had been intended for all stations. It is used by HMCG when there is uncertainty regarding a casualty.

The prefix IMMEDIATE is used for what otherwise might have been PAN-PAN and for when there is apprehension regarding the safety of a vessel.

The prefix FLASH is used in connection with MAYDAY traffic, during the distress phase of an operation, to obtain the quickest possible action when there is reasonable certainty that a vessel is threatened.

Note, however, that these prefixes do not in any way replace the international ones; they are to indicate the urgency of a message between one Coastguard and another. Other prefixes which may be heard are ROUTINE, which is self-explanatory, and FOR INFORMATION, which precedes a message that may have to be written down, although no lifeboat action is required.

Intership working during SAR

It has already been pointed out that a port operation channel becomes a 'net' in the sense that dozens may be listening, although only one or two actually communicate at any one time. The same is true, of course, during a distress working, but in those special circumstances there may be a need for urgent traffic within the 'net'. The main traffic is on Ch. 16 – as already explained – but ships and aircraft engaged in co-ordinated search and rescue might need to communicate with each other, on matters other than would go over the Ch. 16 'net'.

In those special circumstances they may use Ch. 06 – the primary intership frequency. However, it must be stressed that this is a rare condition. Ch. 06 is an intership frequency and HMCG cannot normally use it. Any small craft involved in an SAR incident in any capacity should not change from Ch. 16 unless told to do so by the controlling station: normally HMCG.

Auxiliary Coastguards

As was stressed both in Chapter 1 and Chapter 6, it is HM Coastguard that controls distress traffic on VHF in inshore waters and monitors Ch. 16 continuously. Radio communications have transformed the role of the Coastguard as a co-ordinator and the full-time professional on duty at a Coastguard station increasingly relies on part-time Auxiliaries. The more than 9000 Auxiliaries mostly help to man the various centres or mobile stations (usually Land Rovers). However, several hundred Auxiliary Coastguards are also yachtsmen and they all help to form part of the communications 'net'.

Most Rescue Centres and Sub-Centres will have a group of local yachtsmen or fishermen 'on the books' who act as additional pairs of eyes – in case of an incident – and who can be called upon to help in a search, or to assist in any other manner.

The system is self-correcting to an extent because at weekends or other peak holiday times, when most small craft incidents take place, there is by definition a greater chance that Auxiliaries will be at sea somewhere in the area covered by the Centres.

These reporting Coastguards – called Auxiliaries Afloat – are asked to call the local CG station when they go aboard. They will then pass on the briefest details of their intentions. The centre will log that message and later, if a vessel should be reported overdue, or there was a report from the public of a yacht that appeared to be in difficulties, one of the first checks the CG might make would be to see if any of its Auxiliaries Afloat were in the area and had any knowledge of the craft for which concern was felt.

Auxiliaries Afloat are chosen from those applicants who can show evidence of a wide experience, who have a well-found boat, and who have suitable radio equipment. There is no payment for their service. In certain circumstances Auxiliaries Afloat are granted the necessary Home Office licence to install Ch. 00 – most work on Chs. 16 and 67 – but the principle is the same. The check-in call might be:

THAMES COASTGUARD – THIS IS BARBICAN ON CH. ZERO – OVER.

BARBICAN – THIS IS THAMES COASTGUARD.

THIS IS BARBICAN – ON BOARD RIVER CROUCH – I SHALL BE SAILING FOR BRIGHTLINGSEA ABOUT NOON AND EXPECT TO BE ANCHORED OFF THE STONE FOR THE NIGHT – OVER.

THIS IS THAMES – RECEIVED – OUT.

Note the perfectly acceptable, but abbreviated procedure.

The system works only if the Auxiliary Afloat keeps a continuous listening watch. Thus, for a sailing yacht, a cockpit-mounted loudspeaker is almost essential; otherwise the radio will disturb anyone below who is trying to sleep.

Listening watch: cruising

Chapter 13 outlines a recommendation regarding the keeping of a listening watch by a racing fleet; especially if it is likely to be outside radio range of the normal shore services.

An Auxiliary Afloat will normally keep a listening watch whenever he is on board. It is therefore a natural progression to point out that the more that all yachtsmen on passage do the same, the better the use they will be making of the communications facility.

In some areas where there is a great deal of activity, the constant chatter can become irritating and perhaps this is one of the strongest reasons for stressing the need for discipline (and patience) when using the R/T service. Calling on working frequencies when they are known to be watched will eliminate part of the clutter on Ch. 16. Calls to coast radio stations on working channels (once the change of procedure has been approved) will free Ch. 16 still further. Thus, Ch. 16 will become the calling channel for intership working (except when prior arrangements have been made) and the safety and distress channel: its prime function.

With a cockpit-mounted loudspeaker (and the volume suitably adjusted) it is surprising how quickly the human ear gets used to reacting only to important traffic; it 'switches off' from the rest. A practised communicator will react to:

MAYDAY MAYDAY MAYDAY – the distress call;

MAYDAY RELAY MAYDAY RELAY MAYDAY RELAY – a distress relay;

PAN-PAN PAN-PAN PAN-PAN – an urgency signal;

SÉCURITÉ SÉCURITÉ SÉCURITÉ – a safety signal that precedes a navigation warning;

ALL STATIONS ALL STATIONS – a general call to be read by anyone who intercepts it;

ALL SHIPS ALL SHIPS – a general call that precedes a call to an unknown ship;

Own call sign;

Own name.

It is not necessary to 'switch on your ears' to any other traffic. Except when he is using another channel or monitoring, say, a port operation channel, the professional seaman will monitor Ch. 16 as a matter of course. The amateur, on the other hand, uses his R/T more as a land-line telephone; something he picks up when he wants to use it to speak to somebody.

Even if it does nothing else, I hope this book will encourage yachtsmen to keep a listening watch. (See also Chapter 13: Listening watch while racing.) In that way they become a part of the distress and safety net on which the whole of the VHF communication facility depends.

To transmit to a man who is not switched on is even worse than trying to flash Morse to an astronaut; your procedure, as we have said, might be impeccable but the message will never get through.

PART 2 Code or plain language. Flags, light and sound

Introduction 81

Chapter 9: What type of signal? 82
How bright? — How far apart? — How large? — How loud?

Chapter 10: Methods of signalling 91
The International Code — Flag signalling — Morse Code: Light — Morse Code: Sound — Sound and the Collision Regulations — Signalling by hand flags and arms — Signalling by voice.

Chapter 11: Signalling procedures 103
The Code Book — Flag signalling: Procedure — Use of substitute flags — Code pennant — Flashing light: Procedure — Morse signalling by hand flags or arms: Procedure — Sound signalling: Procedure — Procedural signals — Single-letter signal anomalies.

Chapter 12: Ensigns and special flags 113
Courtesy ensigns — Ensigns when chartering or borrowing — Which ensign? — 'Privileged' ensigns — Where to wear an ensign — House flags — Club burgees — Dressing ship — When to use flags — Requirements of the law: Ensigns and Code Q.

Chapter 13: Signalling and yacht racing 124
Possibility of confusion — International Yacht Racing Union — Yacht race management — Radio for inshore races: UK waters — Radio for offshore racing: UK waters — Radio for offshore racing: Overseas — Radio watchkeeping in the future — Reporting in — Listening watch: Racing.

Chapter 14: Navigation lights and shapes 132
Presence — Aspect — Occupation — Occupations that restrict — Working lights — Small craft lights and shapes — Motor sailing — Anchoring lights and shapes — Summary of sizes.

Introduction

Until comparatively recently, any book on the subject of signalling at sea would have devoted most of its text to the various aspects of the application of the International Code of Signals to flag signalling. In fact there are still books in print (1983) which do just that, even though the present International Code, published in 1969, recognizes that the idea of conversing by the use of flags is a thing of the past.

Today the International Code of Signals concentrates on safety of navigation and of persons. A code is still an essential part of signalling, but the use of the International Code has changed dramatically in the past two decades compared to the practice of the previous two centuries. Nevertheless, there are still many basic signals and signalling methods that have changed little, and the vast majority apply to small craft as much as to the largest bulk-carrier in the world.

Every vessel wears an ensign, in certain circumstances. Every vessel uses a foghorn, hoists anchor lights and carries pyrotechnics. Most commercial harbours have 'entry' signals and nearly all special-purpose vessels have their own methods of indicating their occupation or their presence with lights and shapes. Above all, all vessels exhibit lights at night when they are underway.

Some knowledge and understanding of all these signals is still an essential part of the business of taking a small boat to sea; even if all the 'England expects . . .' type of multiple flag hoist signalling is now a part of history.

9 What type of signal?

It is reasonable to suppose that lights, of one sort or another, have been used on ships since man first learned to control fire. The regular use of flame as a signal certainly goes back as far as the lighthouse off Alexandria in the reign of Ptolemy II, over one hundred years BC. It is all the more surprising, therefore, that the first international agreement regarding the use of lights on the vessel herself was as recent as 1863. Furthermore, it was not until the present generation of navigation lights came onto the market – as a result of the requirements of the 1972 Collision Regulations – that small craft have been able to equip themselves really efficiently.

It is light as a type of *signal* that is of interest here, not the positioning of lights as a means of avoiding collision; but in either sense it helps if a little is understood about what light is and how the intensity affects the range.

Technically speaking, light is an electromagnetic radiation. The amount of that radiation is measured by what is called the luminous flux, the unit for which is the lumen. However, luminous flux, in simple language, is the amount of light emitted in all directions; the amount of illumination. What interests the seaman is the amount of light emitted on one particular direction: the quantity of light that travels from the source of the light to the eye of the man seeing it. This is called the luminous intensity or point brilliance and the unit for measuring it is the candela (cd), often referred to as the candlepower.

How bright?

Throughout the range of intensities that concern the seaman, light is greatly affected by the clarity of the atmosphere and so, before any comparison can be made, this too has to be defined. In earlier versions of the Collision Regulations lights had a range based on what was called 'a clear dark night'. That represented a visibility of 27 nautical miles, which is very rare in northwest Europe, even if it is more common in more tropical climates. Some European countries then suggested working to a standard representing 10 nautical miles. It does not matter all that much what constant is used, but it is obviously essential, in any international agreement, to state precisely how bright a light has to be for it to be seen at so many miles. The current Collision Regulations use a standard agreed by the Inter-Governmental Maritime Consultative Organization in 1972 for a transmissivity factor, as it is called, of $K = 0.8$: this represents a visibility of 13.4 nautical miles.

This chapter is not intended as a thesis on light, any more than the description of the 'capture effect' on VHF R/T, in Chapter 1, amounted to a thesis on radio waves. However, it is necessary to appreciate the bare bones of the problem to understand that twice as bright a light does not shine twice as far, any more than doubling the horsepower of an engine makes a boat go twice as fast.

Figure 4 demonstrates how intensity affects range and it shows that a light of about 1 candela is necessary for a range of 1 nautical mile, a little over 4 candela for 2 miles and 12 candela for 3 miles.

The next fundamental point to understand about light as a signal is that coloured screens filter out a large amount of the available intensity. Even when the screens are of the specially approved types (and the precise specification is included in an annex to the Collision Regulations), most red and green screens filter out between 75 and 80 per cent of the available light.

Thus, bearing in mind that the candela in figure 4 represents the measurement *outside* the lens, a 25-candela light that had had 80 per cent of its intensity absorbed by the filter would show only 5 candela outside the lens, and even that assumes the lens is clean. To put this into more practical language – and remembering that at the wattages we are interested in 1 watt is approximately equal to 1 candela – a 25 W bulb

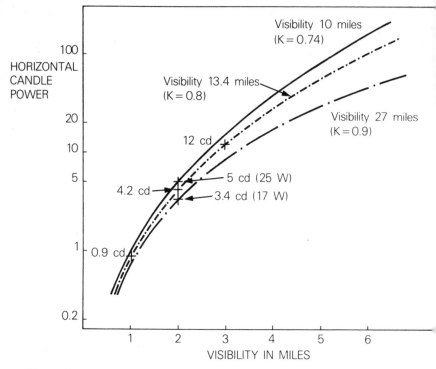

Figure 4

will show about 5 candela outside a clean coloured filter and will thus be visible for slightly over 2 nautical miles.

In theory, and with perfection conditions, a 25 W bulb can be seen at 2.6 nautical miles but, bearing in mind that even a clear glass will filter out at least 5 per cent of the available light, it is now more obvious why 25 watt bulbs are used in most small craft port and starboard navigation lights. For masthead (steaming) or stern lights on the other hand, a 10 W bulb is likely to show appreciably over the required 2 miles. The graph in figure 4 also shows why the white bulb in a big ship's masthead light – which is at least 40 W – is visible for a long time before the ship's sidelights are seen. (See Chapter 14 for small craft navigation lights.)

The three curves in figure 4 show the three different standards used for transmissivity: the 10-mile range used by the Admiralty; the 13.4-mile range used for navigation lights by IMCO; and the old

27-mile range used in the old rules when navigation lanterns were lit by paraffin.

The three curves may not look so very different but, remembering that the vertical scale for the candela is a logarithmic one, look at the candela required for a 2-mile range. It varies, depending on the factor used, from 5 candela to 3.4 candela. Thus, if we are considering coloured light with the screens filtering out at least 80 per cent, those figures for candela outside the lens represent 25 W and 17 W at the source; the higher factor calls for nearly 50 per cent more light to achieve the 'same' range.

At higher powers the difference is even more marked and the main reason why the 10-mile visibility range used by lighthouse and charting authorities was not adopted by IMCO was that to achieve the 6-mile range for masthead lights – required by the new rules for large vessels – the bulb capacities became so great that heat became a problem.

The Admiralty Light List explains the difference between luminous range (the maximum distance that a light can be seen as determined by the intensity of the light and the meteorological conditions), the nominal range (the luminous range when the meteorological visibility is 10 miles) and the relationship with candela.

To give but one example of how the meteorological visibility affects range, assume a visibility of 1 mile instead of the nominal 10 miles. Switch on a light of 100 000 candela. A light of that power would pierce the gloom for 3.2 miles, but even if the light was increased in power to 1 000 000 candela the range would increase by only about 0.6 nautical miles.

How far apart?

Apart from the question of the intensity of a light (and we are talking about what is called point brilliance, not the light flowing out in all directions as from a fluorescent tube) the position of one light in relation to another is also vital. What the scientist calls the 'acuity of the eye' dictates that any two lights (or shapes) must be separated by about one minute of arc to be clearly distinguishable as two lights instead of one. In fact the separation for coloured light has to be slightly greater than for white, but, in practice, one minute, which is just over 0.5 m, at 1 mile is enough. This phenomenon explains why navigations lights have to be separated by a minimum of 1 m in the Collision Regulations. Small craft are allowed, in certain circumstances, to have less brilliant lights than are required for larger vessels, but the separations are never less than 1 m. The same phenomenon also explains why it is comparatively pointless to fit separate port and starboard lights (instead of a combined bi-colour) and then to mount them both on a bow pulpit. Thus mounted they will be only about 0.4 m apart, so that in the only aspect in which they will both be visible at the same time – from right ahead – they will merge into one orange blob when seen from more than half a mile or so. They will not be far enough apart to show as two lights, and thus will defeat the whole point of having two instead of one. (Chapter 14 has more to say about the mounting of lights and of the design of light fittings.)

How large?

The faster the ship the more the likelihood that the officer of the watch or the navigator will spend much of their time with their eyes peering at radar screens instead of finding out what their eyes can tell them from the wing of the bridge, but there are still thousands of small craft where navigation and pilotage are done by eye and for them the brightness of a light or the actual size of a daymark is all important.

A large block of flats or a factory might merely be charted as 'Building conspic.' or 'Chimney' and the precise size is probably unimportant, but for daymarks such as the topmarks of a buoy the size is vital.

As a generalization, a simple shape is clearly distinguishable if the size is between 1:500 and 1:1000 times the distance. The exact proportion depends on the clarity of the atmosphere and even more on the background. In other words, a simple shape such as a ball or a cone can be seen to be a ball or a cone at half a mile, with a clear sky background, if it is approaching 1 m across. Similarly an IALA (International Association of Lighthouse Authorities) buoy topmark, which is rarely less than 1 m across, can be seen to be what it is at least half a mile with the naked eye (figure 5).

It is fairly obvious that a basic shape like an anchor ball can be seen to be 'something' at greater distances than that example and the 3 ft dia. balls that have traditionally been used at the pier head at Dover, for example – where the mariner has to distinguish between two balls or

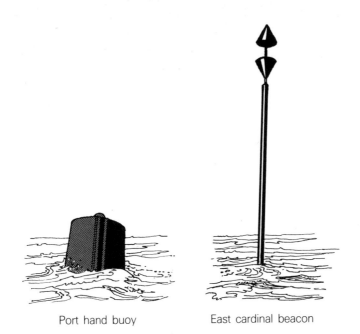

Port hand buoy   East cardinal beacon

Figure 5 *A simple shape is clearly distinguishable for what it is if the size is between 1:500 and 1:1000 times the distance. The exact proportion depends on the nature of the background. Thus a small boat navigator has one advantage over the 'big ship' professional: a topmark is easier to identify when seen from close to the water.*

three balls (1982) – can still be identified at about 1.5 miles. However, for more complicated signals such as those at Zeebrugge, where the mariner has to be able to distinguish between balls and cones point up or cones point down in a three-object hoist, the range is far less.

It is this dilemma that has led to the recent IALA recommendation for ports to use a standard system of lights instead of shapes for port entry signals. As has already been shown, really bright lights are visible even in poor conditions.

To take an extreme example, the leading lights at Hoek van Holland are listed as having a nominal range of 12 nautical miles even in daylight. A shape would need to be at least 25 m across to be clearly distinguishable at that range.

Flags, too, need to be very much larger than might be supposed if they are to be 'read'. The large signal flags used by the navies of the

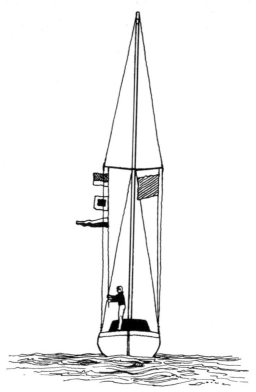

Figure 6 *Code flags of 0·5 m hoist (left) are the absolute minimum for signalling but, as shown here on a yacht about 10 m l.o.a. a single-letter signal which is 1 m in the hoist is by no means too large for an urgent signal or one that has to be read at any distance.*

world – when ships still signalled to each other by flags other than for ceremonial purposes – were at least 6 ft in the hoist and 7 ft 6 in long. At that size the numeral pendant was 15 ft 6 in long!

A small yacht, on the other hand, often uses flags that are merely 1 ft in the hoist and yachting magazines receive a steady trickle of letters to the Editor from 'world-girdling' yachtsmen complaining that their signal 'Please report me to Lloyds', or whatever, was 'ignored' by a passing merchantman.

A flag about 1 m in the hoist and about 1.25 m long is the absolute minimum that is likely to be seen as a signal. It seems enormous when examined in a chandler's shop, but even on a boat about 10 m l.o.a. it seems perfectly reasonable in use. (See figure 6; and also figure 25a and 25b, pages 161–162.)

How loud?

Just as light is affected by the atmospheric conditions, so is sound and it can be a somewhat unreliable method of signalling. Mariners rely on it less than they used to. Even the immensely powerful fog signals used on lightships and offshore oil installations are not nearly as efficient as might be supposed.

Sound travels best in a homogeneous medium. Clear air is good. Fog of even density is quite good, but patchy fog is very bad and sounds can not be relied upon. Sound travels at about 340 m/s through still air and about 1500 m/s in still water, but it is also affected by wind. A very rough approximation of the way in which sound might be heard with or against a fresh to strong breeze is in the ratio of 25:1. This explains why a foghorn on a lightship is sometimes not heard until the ship is quite close to, but the sound can then be heard for a long time afterwards downwind.

To be heard at 1 mile a foghorn needs to produce at least 127 decibels (dB), which is appreciably more than the sound created by the hand-held 'aerosol' type of fog signal usually found on small yachts. The latter, regardless of what some makers claim, can be heard for only about 3 cables in a Force 3 or 4 wind. For comparison the energy needed to generate a sound likely to have a 3.5-mile range is 27 000 W (36 hp).

The frequency of a foghorn is also part of the problem; a very rough indication of the size of the vessel can be obtained by the frequency of her foghorn.

Annex III to the Collision Regulations gives more details on this complex subject, but a small vessel may have a signal from 700 to 250 Hz (approximately an octave and a half from middle C upwards); medium-sized vessels range from 350 Hz down to 130 Hz, and large ships from 200 Hz down to 70 Hz. The lower end of the big-ship scale produces the deep throbbing notes associated with large craft, but it is also possible to combine two distinctly different tones in one signal and achieve a considerably greater range for a lower pressure level.

One of the problems with very loud fog signals is that it is difficult to generate the right volume of sound without experiencing noise levels at open look-out stations that are greater than those considered acceptable for the safety of the crew. However, the problem for small craft is to generate a sound that is loud enough; the problem is never that the sound is too loud. The hand-held 'aerosol' type of horn can be fine for signalling to a bridgekeeper, or for use on inland waters, but it is not adequate for use as a fog signal at sea.

The days when unlimited steam pressure was available for fog signals have long gone. The great advantage of the use of steam for the whistle was that it left a visual trace in the air and, although there was no requirement to do so, the condensed steam was an effective means of saying 'I am the ship making that noise'.

Today sound signals are still important – particularly when several ships may be manoeuvring off a harbour entrance – and the Collision Regulations allow an all-round signalling light to be linked to the whistle signal. It is unlikely that small craft would themselves wish to link signals in this manner, but it is sometimes a great help when large ships are manoeuvring to be able to identify which ship is actually making the signal heard.

For small craft, foghorns can be linked to hand or electric air pumps and, although it is not done as often as it might be, there is a lot to be said for mounting the actual horn aloft. The Collision Regulations call for 'A whistle placed as high as practical . . .' because the range at which a signal can be heard is greatly increased by so doing. Chapter 10 has more to say about the practicalities of signal installations, but the cost of this particular 'modification' is little more than the cost of a length of polythene tube.

10 Methods of signalling

As recently as 30 or 40 years ago – and that is really very recent in the history of international trade at sea – almost all communication between ships was by flags. Flag signalling was a part of maritime life.

When the French Admiral Villeneuve finally decided to send his fleet to sea, at dawn on 19 October 1805, he signalled to his fleet, by flags. HMS *Victory*, cruising about 50 miles off the coast was being kept informed of the movements of the French and Spanish fleets by a chain of small craft passing signals back and forward to each other. *Victory* had received the news of Villeneuve's intentions by 0930 that same morning.

There was no international code in those days. The British used a code devised by Rear-Admiral Popham which used flags and sound signals by day, and sound signals and coloured lights at night. The French and Spanish at that time had a less sophisticated system, and practically no method of keeping in touch with each other at night.

There is strong evidence that the lack of a workable signalling code was a major factor in the final outcome of the encounter.

The International Code

The first international code of signals was the result of a committee set up by the Board of Trade in 1855. It used 18 flags and there were 70 000 different signals. The code was revised in 1887 and, after a conference in Washington in 1889, it was distributed to all the maritime powers.

The First World War upset that version of the code but by 1927 a new one was prepared in seven editorial languages – English, French,

German, Italian, Japanese, Norwegian and Spanish – and by 1932 it had been adopted. There were two volumes, one for visual signalling and the other for wireless telegraphy.

That code lasted, largely unchanged, until the present version, with two more editorial languages – Russian and Greek – was prepared by 1961 and adopted in 1965.

The great difference between the two is that, whereas the earlier code was designed for 'conversation' between ships, today's version caters primarily for situations related to safety of navigation and safety of persons, especially where language difficulties arise. It is suitable for all means of communication, and the principal difference in its design is that today the complicated multiple-hoist messages are a thing of the past: *today every signal has a complete meaning.*

The original 18 flags of the nineteenth-century code were replaced in the version adopted in 1932 by 26 alphabetical flags, 10 numerals, 3 substitute flags and an answering pendant. This same set is still used today. It is not the flags themselves that have been changed, but the manner in which the code book is used.

There are many different methods of signalling. Some, like the Morse Code, are familiar and generally recognized. However, when using the International Code, only certain methods are acceptable. When signalling with sound, there are several single-letter signals that may be used only when in compliance with the requirements of the Collision Regulations.

The methods for signalling by code are:

1 **Flag signalling.**
2 **Flashing light signalling using the Morse Code.**
3 **Sound signalling using the Morse Code.**
4 **Voice; over a loudhailer.**
5 **Wireless telegraphy (W/T).**
6 **Radiotelephony (R/T).**
7 **Morse signalling by hand flags or arms.**

Flag signalling

Any ship of any nationality will use the same set of 26 alphabetical flags, 10 numeral pendants, the answering pendant and 3 substitute flags. To be precise there is also a 'tackline' – a length of halyard to separate two different hoists – but, as a general rule, only one hoist should be hoisted at one time.

When ships regularly made flag signals to each other the Code Book contained detailed instructions regarding the procedure because, with multiple hoists in use, it was essential to recognize the orders in which flags had to be read. Today's code still includes instructions like: 'always hoist the signal where it can be most easily seen, by the receiving station, that is, in such a position that the flags will blow out clear and be free of smoke' – despite the fact that funnels belching smoke are unlikely to be a common hazard! However, it reflects the principle that 'As a general rule only one hoist should be shown at a time' by not even mentioning the order in which a group of hoists has to be read.

For the record, and because multiple hoists might still be employed occasionally, even if only for ceremonial purposes, the order is: masthead, triatic stay, starboard yardarm and port yardarm. When more than one hoist is shown on one halyard they are separated by a tackline, and when more than one hoist is shown on the same yardarm (but on different halyards) the outboard hoist is read first.

As far as small craft are concerned the masthead is the right place from which to fly a club burgee, but any signal flags should be hoisted from the crosstrees (spreaders) and, as tradition has always dictated, the starboard 'yard' is read first. Today it is rare to find more than two signal halyards – one on each spreader. However, although the racing yachtsman today often uses a backstay for a class flag or pennant, a sea-going yacht will be embarrassed by having less than two halyards (one port and one starboard) because there are occasions when one is needed for a courtesy ensign (a small version of the maritime flag of the country being visited) and the other for a code flag Q, for Customs clearance. (See also Chapter 12.)

Apart from such signals (a courtesy ensign is always worn to starboard because an ensign takes precedence over any single flag), it must be admitted that the occasions when code flags are *needed* are very rare indeed. However, that does not alter the fact that the system is international and one day, your own or somebody else's life might be saved by a knowledge of the basic system of signalling: by flags as well as by other methods.

Morse Code: Light

Until comparatively recently Morse Code by flashing light was a common part of sea-going life (see figure 7). Occasionally we still see emotive pictures of a seaman standing by Morse searchlights, complete with the clattering, venetian-blind type of shutter across the front. In the real world of today, ships do not have signalling searchlights on the wings of the bridge, but they do all carry the so-called daylight signalling lamp, better known as an Aldis lamp.

In their cruder versions – and some are still available in ex-government 'junk' shops – the press-to-operate switch in the handle was an electric switch so that the bulb was actually turned on and off. However, this method had obvious limitations for the professional signaller because of the time taken for the light to reach its full, intense brightness and then dim down again. It was replaced by a rocking device in the focusing mirror so that the concentrated beam was itself moved up and back again for each dot or dash in the code. This still rather crude solution was then replaced with the type used today where a matt-black tubular shield masks the bulb. The light stays on all the time the lamp is in use, and it is the rays from the bulb that are masked.

To achieve the 'daylight' effect in the modern signalling lamp – and it is bright enough to be used by day – the rays from the bulb are concentrated into a very narrow beam. Unfortunately this means that the modern Aldis is of very little use to the small craft signaller because, except in really calm weather, it is impossible to keep the narrow beam trained on the eye of the man receiving the signal. The much cruder lamps dating from the 1940s, which give a considerably wider and less concentrated beam, are somewhat better although they are of little use by day.

For those who enjoy searching through the few ex-government surplus shops that still exist, there was one signalling lamp, made from light, spun brass and with hand holds on each side, which was for use on aircraft. It had a beam width of about 40 degrees, and was therefore far more suitable for small craft use but, alas, the days of cheap 'junk' have largely gone.

Today there are several so-called spot lights available, and the better versions have a thumb switch in the handle so that they can be used for slow signalling, if there is not too much motion. A much more

A	.—	J	.———	S	...	2	..———
B	—...	K	—.—	T	—	3	...——
C	—.—.	L	.—..	U	..—	4	....—
D	—..	M	——	V	...—	5	.....
E	.	N	—.	W	.——	6	—....
F	..—.	O	———	X	—..—	7	——...
G	——.	P	.——.	Y	—.——	8	———..
H	....	Q	——.—	Z	——..	9	————.
I	..	R	.—.	1	.————	0	—————

A dash — is equal to three dots ...
The space between parts of the same letter is equal to one dot.
The space between two letters is equal to three dots.
The space between two words is equal to seven dots.

Ä (German) = AE (Danish)   .—.—   (RT)
Á or Å (Spanish or Scandinavian)   .——.—
Ch (German or Spanish)   ————   (OT)
É (French)   ..—..
Ñ (Spanish)   ——.——
Ö (German) = Ø (Danish)   ———.   (OE)
Ü (German)   ..——   (UT)

The following letters more likely to be understood by the occasional signaller if sent as follows:

Ü (German) = UE   Ä (German) = AE
Å (Norwegian) = AA   Ö (German) = OE
Ø (Danish) = OE

They are not used as a part of the International Code of Signals, but they are used, locally, in the spelling of names.

Those who do not normally use accented letters should note that although an accent in French is basically a stress mark – it affects the way a word is spoken – a letter with an accent in some Scandinavian languages can stand for a different letter of the alphabet. In Norway, for example, words beginning with Å or Ø appear in an alphabetical list after Z.

Figure 7 *International Morse Code*

satisfactory answer for those who want a Morse signalling facility that is likely to be efficient in all conditions is to wire an all-round masthead-mounted white light to a Morse key.

Chapter 11 describes the Morse procedure signals, including those that apply to other methods of signalling, but when signalling by flashing light the rhythm is very important. A dot is taken as *the* unit. A dash is equivalent to *three* units. The space of time between any two elements of a symbol is equivalent to *one* unit; between two complete symbols equivalent to *three* units; and between two words or groups equivalent to *seven* units.

Morse Code: Sound

In the hands of a professional, light signalling can be almost unbelievably fast. Sound signalling, on the other hand, is necessarily slow. The International Code book recommends that, except in an emergency, sound signals other than single letters should be avoided.

In the full list of the International Code single-letter signals, there are ten marked with an asterisk, which signifies that they may be made by sound only when in compliance with the Collision Regulations.

Under the heading 'How loud?' on page 89, the advantage of having a foghorn mounted aloft was stressed. However, from the close-quarters signalling point of view, it is the immediate availability of the signal, rather than its volume, that is the more important.

In practice no one wants the complication of two sound-signalling systems and thus the two, somewhat different requirements – volume of sound for a fog signal, and the immediate availability and clarity of sound as a signal – have to be combined.

Sound and the Collision Regulations

When sound signals are used in accordance with the Collision Regulations the 'short blast' is of about one second's duration, the 'prolonged blast' is from four to six seconds and there is an interval of about two seconds between blasts. The Collision Regulations refer to a 'whistle' but that is taken as meaning any sound signal capable of producing the prescribed blasts.

The ten single-letter signals that may be made by sound only when in compliance with the Collision Regulations are: B, C, D, E, G, H, I, S, T and Z. Some have the same meaning in the International Code of Signals as when used as a signal from the Collision Regulations, but others do not. Z, to give but one example, means *I require a tug* when made with a flag or by flashing light, but as a sound signal it is a part of Rule 34 and means *I intend to overtake you on your port side*.

Although the professional is required to have a very full knowledge of all sections of the Collision Regulations, it would be a little optimistic to assume that all amateurs will have the same grasp. Nevertheless, it is important that all the common sound signals should become familiar and it is best if they are learned as a group. The fact that E has the same meaning when signalled by flag, light or sound, although Z does not, is unimportant. What matters is that sound signals are recognized.

The commonest – and not unnaturally the most useful – are:
E – one short blast: *I am altering my course to starboard*.
I – two short blasts: *I am altering my course to port*.
S – three short blasts: *I am operating astern propulsion*.

They are heard frequently when made by commercial ships – supplemented on some vessels with an all-round flashing light while the manoeuvre is being carried out [Rule 34(b)] – and they are also the signals most likely to be needed by small craft. There are numerous occasions when small vessels will be milling about in close quarters – while approaching a lock or marina entrance for example – and the ability to indicate one's intentions or to receive similar information from another vessel can be invaluable.

With one exception – the prolonged blast from Rule 34(e) – the sound signals referred to here do not have any 'Look out, here I come' meaning as may a motorcar horn. They have a precise meaning.

In addition to the one, two and three short blasts there are three other single-letter manoeuvring signals, but two of them have meanings that are completely different to the significance of the letters when exhibited as flags. Two prolonged blasts followed by one short blast (Code G) means: *I intend to overtake you on your starboard side*, and two prolonged blasts followed by two short blasts (Code Z) means: *I intend to overtake on your port side*. The third of these signals is Code C – one prolonged, one short, one prolonged and one short – which is used by a vessel indicating agreement to the overtaking signal she has heard; but as Code C means *Yes*, its use for this purpose is quite logical.

All these manoeuvring signals are contained in Rule 34 and it includes two more that have no connection at all with Morse Code. This supports the suggestion that the sound signals in the Collision Regulations should be looked upon as signals in their own right, rather than as items from the International Code of Signals; nowhere in the Collision Regulations is the term 'Morse Code' mentioned.

Nevertheless, the commonest of the sound signals in the Collision Regulations do have a dual role and Code D – *Keep clear of me I am manoeuvring with difficulty*, and Code M – *My vessel is stopped and making no way through the water*, have precisely the same meaning (even if the wording is not exactly the same) in Rule 35 'Sound Signals in Restricted Visibility' and in the single-letter section of the International Code. (See figure 8 for full details of the sound signals from the Collision Regulations and see also the footnote on page 109.)

Figure 8 *Sound Signals and the Collision Regulations*

Manoeuvering and warning signals (Rule 34)
A short blast is one of about one second duration
A prolonged blast is one of from four to six seconds duration.

When vessels are in sight of one another:
—   Altering course to starboard
— —   Altering course to port
— — —   operating astern propulsion

Those signals may be supplemented, if required, by light signals which may be kept flashing whilst the manoeuvre is being carried out. The flash and the interval between flashes is about one second; the interval between successive signals is not less than ten seconds.

In a narrow channel a vessel intending to overtake may signal:
——— ——— —   Intend to overtake on your starboard side.

——— ——— — —   Intend to overtake on your port side.

The vessel about to be overtaken signifies agreement with:
——— — ——— —

When there is doubt whether sufficient action is being taken by another vessel to avoid collision, make:
— — — — — —   (at least five short and rapid blasts)

Nearing a bend where another vessel may be obscured make:
———
which may be answered by a similar prolonged blast by any approaching vessel.

Restricted visibility (Rule 35)
———   Power driven making way
———   ———   Power driven underway but stopped
———   ———   —   Not under command, restricted in ability to manoeuvre, constrained by her draught, a sailing vessel or a vessel towing.
———   —   —   —   A vessel towed (where possible made immediately after the signal from the towing vessel.)

🔔 Vessel at anchor (Bell, for about 5 seconds)
—   ———   —   *Additional* signal which may be sounded by vessel at anchor to give warning of her position to an approaching vessel.
—   —   —   —   Pilot vessel, identity signal *in addition* to other sound signals.

Signals to attract attention
A continuous sounding with any fog signalling apparatus is an international distress signal.
—   —   ———   International Code "U" – *You are running into danger* – is a useful single-letter sound signal to memorize.

Port signals
Many ports have by-laws which make special provision for their own requirements. Several major ports use the following:
—   —   —   —   —   I am turning right round to starboard
—   —   —   —   —   —   I am turning right round to port
It is a signal sufficiently well established to commit to memory. Finally, several of our neighbours use:
———   ———   ———   on their inland waterways, as a signal to a bridge attendant. Technically it conflicts with the single-letter code "O" – *Man overboard* – but it is used throughout the canal system of Holland, nevertheless.

Finally, after stressing the advantages of having the facility for sound signalling aboard small craft, it must be pointed out that the 'starboard', 'port' or 'astern' signals which are so common as well as useful, apply only when under power.

The requirements for the sound signals in restricted visibility in Rule 35 – Morse D for a vessel under sail – are mandatory for vessels of 12 m l.o.a. or over; vessels of less than 12 m are required to 'make some other efficient sound signal'.

The other two types of sound signal referred to in the Collision Regulations are the gong and the bell. The former applies to vessels at anchor that are over 100 m l.o.a., or to vessels that are aground, and therefore the requirements to use a gong do not apply to small craft. The bell, on the other hand, is required for any vessel at anchor of more than 12 m l.o.a. and the traditional fisherman's signal of a spanner banged back and forwards in a saucepan is not really a satisfactory alternative.

It would be optimistic to suggest that yachtsmen normally use a bell when at anchor in restricted visibility, but it is reassuring to have the ability to make 'a suitable signal' and a bell, rung vigorously, makes a highly satisfactory noise.

Small craft, because of the violence of the motion on occasion, would not normally keep a bell permanently rigged. Thus, a portable bell, to be rigged if visibility is bad, is the alternative. However, bells of less than about 20 cm in diameter are liable to sound somewhat tinny. (See also the reference to sound signals when at anchor in Chapter 11). Code R made by sound is a signal meaning much the same as the bell. Those who do not carry a bell might use the foghorn, when at anchor, to sound R. Although it might not be understood by all, it would certainly qualify as 'some other efficient signal' for the smaller vessel.

Signalling by hand flags or arms

Since 1976, as far as the United Kingdom is concerned, semaphore as a means of communicating has been considered obsolete. In one sense that is a great pity. Up until the 1950s semaphore with hand flags was a normal part of intership communication in harbour, or even at sea when at close quarters. It was not difficult to learn and it needed no special equipment. However, once the radiotelephone began to be looked upon as the norm, semaphore was little used and, if not used, it was forgotten.

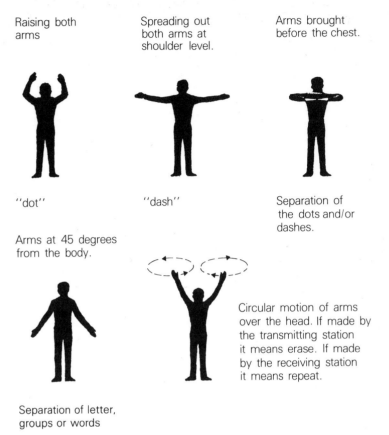

Figure 9 *Signalling Morse.*

In 1976 the decision was made to stop teaching it. So semaphore is now dead.

Signalling Morse, however, by flags or arms (figure 9) is retained in the Code Book as an internationally recognized method and the fact that it is very rarely used need not detract from its efficiency.

Imagine the problems for a man whose boat has stranded on an offshore bank and a passing yacht notices his dilemma and heaves to in order to try to find out what is happening. The yacht might be in a position to signal by several methods, but the casualty may have lost everything except his voice (which may well be almost useless) and his arms. The ability to pass a message by Morse Code – even at a very slow rate – might be the only way to allow him to communicate. The fact that a man is standing on a sandbank beside his grounded yacht does not necessarily mean he is in distress, but he may well be in need of assistance and even a basic knowledge of Morse could enable him to outline the nature of that assistance.

Signalling by voice

In the same category as the evocative picture of the signalman standing with his hand on the lever that operates the shutter of a signalling searchlight, there is the image of the captain of a square-rigged ship standing on the rain-swept poop deck and bellowing to his men aloft through a huge megaphone. Today, however, the mate of the watch is likely to be inside the wheelhouse. If he wants to speak to the forecastle head he does it via an intercom with a loudspeaker, and if he wants to speak to another vessel or to the pilot boat he uses VHF, not a loudhailer.

For small craft, too, the loudhailer is rarely used and for similar reasons. A dock entrance is likely to have 'traffic signals' to tell a man when to stand off and when to enter, and increasingly yacht harbours rely on the radiotelephone rather than the megaphone.

If the voice is used, whether amplified or not, the message should be as simple as possible.

As a generalization, shouting is to be discouraged. It is often ineffective and it invariably irritates others. However, a well-timed shout of 'Look ahead, Sir' (or whatever might be appropriate) can serve a very useful purpose in such circumstances as when a dinghy tacks, without looking, right under your bow.

In addition, there are several universally recognized hand or body signals such as beckoning or pointing which often replace or supplement the shout. (See also figure 26, page 173.)

11 Signalling procedures

Until recently the business of learning the procedure for flag signalling was a major part of the whole subject of communication because, while they were still used for 'conversation', there were long series of flag hoists to be prepared, read, and acknowledged before they could be answered. Today, although procedure is still of vital importance, there is very much less that needs to be committed to memory: the 'every signal has a complete meaning' philosophy has simplified the whole process.

The code book

There is not the slightest need to try to learn the International Code off by heart – in fact it would be impossible. What is important is to learn the basic principles. Regarding the code itself there is no short cut: you buy the book – and there are three Amendments (1982) that apply to the basic 1969 Code.

The principle is simplicity itself. Single-letter signals are allocated to very important, urgent or common-use matters. Everything else (except for the medical section) is a two-letter signal. The single-letter meanings are not difficult to memorize, but it is by no means essential to do so. Remember it is the principle of the code that matters; the meanings can usually be looked up in the code book without too much delay.

For example, both single- and two-letter signals can be qualified by one of the three complements tables in the code book.

A   *I have a diver down; keep well clear at slow speed.*
K   *I wish to communicate with you.*
Y   *I am dragging my anchor.*

are examples of basic single-letter messages.

DJ   *Do you require a boat?*
GU   *It is not safe to fire a rocket.*
UH   *Can you lead me into port?*

are examples of basic two-letter messages.

The 'complements' qualify messages in different styles. Some signals have a number of complements. Thus:

KR   *All is ready for towing.*
KR 1   *I am commencing to tow.*
KR 2   *You should commence towing.*

This type of complement is listed, under a subject heading, in the text of the code, but some single-letter signals have complements to give special meanings.

A with three numerals	Azimuth or bearing
C with three numerals	Course
D with two, four or six numerals	Date
G with four or five numerals	Longitude (the last two numerals denote minutes, the rest degrees)
K with one numeral	*I wish to communicate with you by . . .* (Complements table I)
L with four numerals	Latitude (the first two numerals denote degrees and the rest minutes)
R with one or more numerals	Distance in nautical miles
S with one or more numerals	Speed in knots
T with four numerals	Local time. The first two numerals denote hours and the rest minutes.
V with one or more numerals	Speed in kilometres per hour
Z with four numerals	GMT The first two numerals denote hours and the rest minutes.

All these signals may be made by any method of signalling but the 'odd-man-out' is letter **K**, for which there is a different type of complement.

Tables of complements

There are three tables of complements, but they are used only as and when specified in the text of the signal.

Apart from **K** (the example just mentioned), Table I applies to signals like:

YR *Can you communicate by* . . . (Complements table I)
YT *I cannot read your* . . . (Complements table I).

Table I 1 Semaphore
2 Morse signalling by hand flags or arms
3 Loudhailer (megaphone)
4 Morse signalling lamp
5 Sound signals
6 International Code flags
7 Radiotelegraphy 500 kHz
8 Radiotelephony 2182 kHz
9 VHF Radiotelephony – channel 16

Table II applies to a signal like:
TZ *Can you offer assistance?* (Complements table II)

Table II 0 Water
1 Provisions
2 Fuel
3 Pumping equipment
4 Fire-fighting appliances
5 Medical assistance
6 Towing
7 Survival craft
8 Vessel to stand by
9 Ice-breaker

Table III applies to a signal like:
NB *There is fishing gear in the direction you are heading* (*or in the direction indicated* – Complements table III)

Table III   0 Direction unknown (or calm)
              1 Northeast
              2 East
              3 Southeast
              4 South
              5 Southwest
              6 West
              7 Northwest
              8 North
              9 All directions (or confused or variable)

Thus, with this clever use of complements tables:

K 9 means *I wish to communicate with you by VHF R/T Ch. 16*
TZ 4 means *Can you offer assistance with fire-fighting appliances?*

and

NB 6 means *There is fishing gear to the west of your position.*

This brief summary of how the code book is arranged shows how signals are composed and how a basic message can be elaborated by the use of complements. It also demonstrates how the code is arranged so that, as far as possible, each signal has a complete meaning.

Everything, so far, has applied to all methods of signalling with the International Code. However, for obvious reasons, the detailed procedures vary according to the method of signalling that is used.

Flag signalling: Procedure

As a general rule only one hoist should be shown at a time, and if more than one group is shown on the same halyard (a condition unlikely ever to apply aboard a small yacht because of the lack of actual height for the two groups) the two must be separated by a tackline.

If it necessary to signal a particular vessel her identity should be included in the first hoist, and the method of doing so is described under the heading 'Communications' in the International Code book. However, a signal that does not have an identity is understood to apply to all stations within visual signalling distance, and all the single-flag hoists that are likely to be used by small craft fall into this second category.

To answer a signal, the receiving station hoists the answering pennant at the dip (about half way), and then close up (hoisted to the full extent of the halyard) immediately the transmitting station's hoist is understood, returning the answering pennant to the dip when the transmitting station hauls down her hoist.

There are special signals in the code book to ask a station to identify herself, as well as for use if a signal is not understood, but these are both examples of codes that do not need to be committed to memory.

What does have to be memorized is the use of the substitute flags. They are an essential part of the use of flags and once understood, the principle will never be forgotten.

Use of substitute flags

The three substitute flags allow virtually any signal to be made with only one set of flags on board.

The first substitute always repeats the uppermost flag *of that class of flag that immediately precedes the substitute*. The second repeats the second, and the third the third, but always counting from the top of that class of flag which immediately precedes the substitute. Thus, no substitute can ever be used more than once in the same group and the answering pennant (used as a decimal point when one is necessary) is disregarded when determining which substitute to use.

Examples:

VV would be made by flag V – first substitute.

The number 1100 by numeral 1 – first substitute – numeral 0 – third substitute.

The signal L2330 would be flag L – numeral 2 – numeral 3 – second substitute – numeral 0. Note that it is the *second* substitute that is used because although the numeral 3 is the third flag in the hoist, it is the second numeral.

Code pendant Finally, note that if a ship of war wishes to communicate with a merchant vessel, she will hoist an answering pendant (called for this purpose the code pendant) in a conspicuous position and keep it flying during the whole of the time that the signal is being made. The reason is that ships of war (as they are called in the code book) do not normally use the International Code: hoisting the answering pendant indicates that she is using the International Code, not a private one.

Flashing light: Procedure

As already mentioned, the professional (such as a naval yeoman of signals) can use a lamp in a manner that appears almost unbelievable. Small craft, on the other hand, are likely to use light for signalling only in an emergency. Nevertheless, the procedure is exactly the same and it is important for the amateur to remember that it is far easier to learn to send than to receive. In consequence there is a common tendency to send too fast. The receiving station will (or should) reply at the speed of the sender's traffic, and if that is faster than the sender can receive there is bound to be delay.

A signal is divided into: a **call**; the **identity**; the **text**; and the **ending**.

The call: $\overline{AA}$ $\overline{AA}$ $\overline{AA}$ $\overline{AA}$ (continued as necessary) is the call to an unknown station or a general call to attract the attention of all stations within signalling distance. It is continued until the station addressed answers.

The answering signal: A call is answered by $\overline{TTTTTTTT}$, which is continued until the calling station ceases to transmit. The calling station then makes DE (which means 'from' by all appropriate methods of signalling) and that is followed by the name or call sign of the station calling. The letter T is used to indicate the receipt of each word or group.

The text: The calling station then sends her message, each word or group being acknowledged by T.

If she makes a mistake the station sending makes $\overline{EEEEEE}$, to indicate that the last word or group was incorrect. When answered, the transmitting station will repeat the last word signalled correctly and then continue. There are also special procedural signals to allow either station to ask for a repetition and the text ends with $\overline{AR}$.

The ending: $\overline{AR}$ is the signal for the end of a transmission. It is answered with R. Note that R – 'Received' or 'I have received your last signal' is also used as a procedural signal with the same meaning in radiotelephony: 'Received' in voice – or 'Romeo' in voice for the letter R.*

When signalling, a bar over a group of letters signifies that they are to be sent as one symbol: $\overline{AAA}$, for example, is used when necessary to signify a decimal point or a full stop.

Repetition

RPT is used by the transmitting station to indicate that she is about to repeat, but if such repetition does not follow it is a request for the receiving station to repeat what she has just received. A correctly received repetition is acknowledged by OK.

(See also pages 110, 111 for procedural symbols that apply to Morse signalling other than by light.)

Morse signalling by hand flags or arms: Procedure

Now that semaphore has been abandoned, the use of flags or arms for signalling Morse is the only other technique that might be of use for small boats. The fact that it is extremely rarely employed need not detract from its simplicity and therefore its potential usefulness in an emergency. The code book shows illustrations of smartly dressed sailors in uniform caps with signal flags. In practice I suggest that a small craft which had no daylight signalling lamp, no code book and no radio might still send a vital message – even if very slowly – by Morse signalling by arms.

The procedure is really the same as for Morse by lamp. K2 – *I wish to*

---

*It is often easier to understand, and therefore remember, a signal once its origin has been traced. Rule 35(f), *Sound Signals in Restricted Visibility*, refers to the use of three blasts in succession, one short, one prolonged and one short (Morse R) for use by a vessel at anchor which, in addition to her bell, may 'give warning of her position and of the possibility of collision to an approaching vessel.' Under the pre-1965 rules that signal was referred to as one that a vessel at anchor might give if she heard another approaching. It meant 'You may feel your way past me'. In other words it was an answer/acknowledgement of having heard the signal of a vessel making way. Today it has a slightly different meaning but is still, in effect, an answering signal meaning: 'I have received a signal from an approaching vessel and I wish to attract her attention.'

communicate with you by hand flags or arms – is the code for this method of signalling, but $\overline{AA}\ \overline{AA}\ \overline{AA}$ may be used instead. $\overline{AA}\ \overline{AA}\ \overline{AA}$ and T should be used respectively by the calling station and by the station called. Normally both arms would be used – but note that one important advantage of this method of signalling, compared with the now obsolete semaphore, is that it can still be used with only one arm.

All signals end with AR.

Sound signalling: Procedure

As already explained under the heading 'Methods of signalling', sound signalling is slow and easily causes confusion. Anything other than single-letter signals should be used only in emergency. If used, the signals must be made slowly and at sufficiently long intervals to ensure that no confusion can arise. (See also Chapter 10: Sound and the Collision Regulations and figure 9, page 101.)

Procedural signals

The following are for use where appropriate in all forms of transmission:

AA   'All after . . .' (used after the Repeat signal – RPT), means 'Repeat all after . . .'.
AB   'All before . . .' (used after the Repeat signal – RPT), means 'Repeat all before . . .'.
AR   Ending signal or end of transmission or signal.
AS   Waiting signal or period.
BN   'All between . . . and . . .' (used after the Repeat signal – RPT), means 'Repeat all between . . . and . . .'.
C   Affirmative – YES or 'The significance of the previous group should be read in the affirmative'.
CS   'What is the name or identity of your vessel (or station)?'
DE   'From . . .' (used to precede the name or identity signal of the calling station).
K   'I wish to communicate with you' or 'Invitation to transmit'.
NO   Negative – NO or 'The significance of the previous group should be read in the negative'. When used in voice transmission the pronunciation should be 'No'.

OK Acknowledging a correct repetition or 'It is correct'.
RQ Interrogative, or 'The significance of the previous group should be read as a question'.
R 'Received' or 'I have received your last signal'.
RPT Repeat signal 'I repeat' or 'Repeat what you have sent' or 'Repeat what you have received'.
WA 'Word or group after . . .' (used after the Repeat signal – RPT), means 'Repeat word or group after . . .'.
WB 'Word or group before . . .' (used after the Repeat signal – RPT), means 'Repeat word or group before . . .'.

Notes:
1 The procedure signals 'C', 'NO' and 'RQ' cannot be used in conjunction with single-letter signals.
2 When these signals are used by voice transmission the letters should be pronounced in accordance with the letter spelling table, with the exception of 'NO' which in voice transmission should be pronounced as 'No'.

Single letter signals

A I have a diver down; keep well clear at slow speed.
*B I am taking in, or discharging, or carrying dangerous goods.
*C Yes (affirmative or 'The significance of the previous group should be read in the affirmative').
*D Keep well clear of me; I am manoeuvering with difficulty.
*E I am altering my course to starboard.
F I am disabled; communicate with me.
*G I require a pilot.
*H I have a pilot on board.
*I I am altering my course to port.
J I am on fire and have a dangerous cargo on board: keep well clear of me.
K I wish to communicate with you.
L You should stop your vessel instantly.
M My vessel is stopped and making no way through the water.
N No (negative or 'The significance of the previous group should be read in the negative').
O Man overboard.
[P I require a pilot. (Sound only. See page 112)
Q My vessel is 'healthy' and I request free pratique.
*S I am operating astern propulsion.
*T Keep clear of me: I am engaged in pair trawling.
U You are running into danger.
V I require assistance.
W I require medical assistance.
X Stop carrying out your intentions and watch for my signals.
Y I am dragging my anchor.
*Z I require a tug.

Signals marked * may be made by sound only in compliance with the Collision Regulations.

Single-letter signal anomalies

Most professional seamen will memorize the meanings of all the single-letter signals but, as has already been mentioned, the occasional sailor does not have quite the same need.

There are a few anomalies and duplications in the list. For example, Code G, P, and Z all have a second, alternative, meaning when used by fishing vessels operating in close proximity. I would be surprised to hear if they are ever used today, because fishermen seem to keep a VHF channel open when they are in a group, but special signals concerning nets being hauled, becoming fast or being shot do exist.

Flag N is somewhat non-standard in that it may be used only visually or by sound. For voice transmission the International Code states that 'No' should be used; however, the Services and HM Coastguard use the word 'Negative' in voice.

Finally, K and S have a special significance as distress messages.

Observant readers will have noticed that R is missing from the otherwise full alphabetical list of single-letter signals (see page 111). Letter R, in the International Code, means 'Received' or 'I have received your last signal'. However, when signalling with flags the answering pennant is used to indicate that a hoist is seen and then understood. With Morse the answering signal is T, to indicate the receipt of a word, and R to acknowledge the end of a message.

In radiotelephony the phonetic 'Romeo' is correct if there are any language difficulties, otherwise 'Received' is the proper acknowledgement.

Clearly, R as a single-letter code signal has no logical place, and as the Morse _ _ _ _ _ in sound also has a special meaning (as has just been explained), R is omitted from the single-letter list.

Finally, the UK Safety of Navigation Committee recommended to IMO in 1982 that, because the International Code flag signal **G** – *I require a pilot*, can no longer be made by sound (because it conflicts with the 1972 Collision Regulations sound signal for a vessel which intends to overtake another to starboard) the International Code signal **P** should be used *'at sea'* as a sound signal to mean – *I require a pilot*.

This recommendation is likely to be accepted by the IMO Marine Safety Committee because, despite the extensive use of VHF R/T, it is obvious that there is still a need, on occasion, to summon a pilot with a sound signal.

12 Ensigns and special flags

The decline in the use of flags for communication has been caused by the huge technological advances in the use of radio, and it is possible that this may have accelerated the change in nearly every seaman's attitude to ensigns.

Once upon a time the ensign was a key factor about a ship. It showed her nationality – was she friend or foe? Nowadays many ships fly flags that represent the nationality of the company which owns them (so-called flags of convenience), but the shares in that company may well be owned by nationals of several other countries. However, the fact remains that the use of a national ensign (even a flag of convenience) is still governed by statutory regulations.

In Britain it is the Merchant Shipping Act that applies, but there are also customs and traditions affecting the use of ensigns and it can cause considerable offence to ignore them.

Books on the subject of signalling usually devote pages, if not chapters, to the etiquette associated with ensigns and, no doubt, in the days when yachts were comparatively few and far between (and they nearly all had paid crew anyway), the owners took a delight in watching to see if their ensigns were being hoisted and lowered at a precise and appropriate moment.

Rightly or wrongly, that custom is now dead. There are occasions – during an important regatta for instance – when a few yachts will still follow tradition. But with signalling customs changing as fast as they are, it seems far more sensible to try to maintain the few customs that still mean something, rather than to write about what the paid crew should do if the owner is ashore at a particular time of day, or the precise order of flags that should be used when 'dressing ship'.

The Merchant Shipping Act (parts of which were under review in 1982) requires that an ensign is worn at sea 'if there is sufficient light for it to be seen . . . on falling in with any other ship, or ships, at sea or when within sight of or near land and especially when passing or approaching forts, batteries, signal or coastguard stations, lighthouses or towns.'

It is obvious from the wording of the Act that it was composed for a somewhat different era. However, it is required of any vessel over 50 tons to wear her colours (her national flag) when entering or leaving any British port, and it is also mandatory for all vessels – except registered fishing vessels – to wear their colours when entering or leaving any foreign port.

Courtesy ensigns

The requirements for ensigns are mandatory, and understandably so, but the requirements for courtesy ensigns – a smaller version of the maritime flag of the country being visited – are purely a matter of custom.

In the United Kingdom we are not particularly flag-conscious. Among our near neighbours, on the other hand, some people place very much more importance on the meaning of, and traditions associated with, a national flag. In the Netherlands and in the Scandinavian countries in particular, very considerable and quite needless offence can be caused by a casual or off-hand attitude.

In Chapter 10 it was explained that, in the days when flag signalling was still the norm and when multiple hoists were still used, there had to be an internationally agreed order in which the hoists were read. From that tradition has come the modern custom of wearing a courtesy ensign in the *starboard* rigging; it was always the starboard hoist that had to be read before the port.

Ensigns when chartering or borrowing

Confusion can arise when chartering. There might appear to be logic in recommending that the ensign that a yacht wears should always represent the country of the man in charge. However, that philosophy is impossible because of the laws in most – although not all – countries.

If a yacht is registered in the UK, in the sense that she has a Certificate of Registry, she becomes a registered British ship. So, even if she is being sailed by a Dane in German waters, she still wears a British ensign. However, the problem is even more complicated because of what has just been said regarding courtesy ensigns. If our mythical British yacht was being sailed by a Danish friend of the owner in West German waters, she would be expected to wear a West German courtesy flag. There could easily be a situation where a national of country A was sailing in the waters of country B, with a B courtesy flag in his starboard rigging, and an ensign from country C on the stern: nothing, however, to represent his own nationality!

In an attempt to make some sense of this anomaly, it is generally agreed that it would be perfectly correct for the skipper of a chartered or borrowed yacht to wear a small version of the maritime flag of his own country in the port rigging.

There is no historical precedent for the suggestion, but it was first made several years ago, following queries to the Royal Yachting Association. I, among others, was consulted in my role as the then chairman of the RYA Cruising Committee. The idea was first published in 1979 in the RYA *Flags and Signals* booklet.

Thus it is recommended that our mythical Dane might sail a British yacht with a British ensign (showing that she was either a Registered British ship or owned by a Briton) with a Danish maritime flag in his port rigging, and the courtesy ensign of the country he was visiting in his starboard rigging.

In view of the large amount of chartering that is now done in some parts of northwest Europe, particularly in the Baltic, the idea has merit. It would at the very least avoid the somewhat silly situation, which I once experienced in Danish waters, when the yacht lying alongside my own had a Swedish ensign on the stern and a Danish courtesy flag in her rigging, but the charterers did not speak one word of either of those languages.

Which ensign?

The whole subject of signalling with flags, and the emotive meanings associated with 'striking the colours' in battle and so on, are so deeply engrained in maritime history that it is not surprising that some countries have more than one national flag. Those interested in the history of the subject cannot do better than to read *Flags of the World* revised by Captain E. M. C. Barraclough, 1965, Frederick Warne & Co.

An ensign is a maritime flag and some countries, Britain for example, not only have different national flags, but they have more than one maritime ensign.

The national flag in Britain is the Union Flag (usually, but incorrectly, called the Union Jack). The navies of the world tend to use different ensigns – the White Ensign in the British navy – and we complicate the whole business still further by having 'privileged ensigns' as well.

The British maritime ensign is the Red Ensign. A registered yacht is entitled, under the Merchant Shipping Act, to wear the Red Ensign as her national colours and, although not entitled by statute, it has been generally accepted for a great many years that an unregistered yacht should also wear the Red Ensign when it is proper for her to show her national colours.

From this it can be seen that it is a small version of the Red Ensign, never the Union Flag, which it is correct as a courtesy flag on visiting foreign vessels.

The confusion arises because of the fact that, for reasons steeped in history, numerous British yacht clubs were given warrants to use the undefaced blue, the defaced blue or the defaced red ensigns, respectively, as a privilege.

'Privileged ensigns'

The law in the United Kingdom concerning special ensigns is complicated and enshrined in Section 73 of the 1894 Merchant Shipping Act. Permission is granted only to certain registered vessels and under certain circumstances and the warrant, issued by the Ministry of Defence, must always be carried on board if a special ensign is being worn.

In practice, and because of the enormous growth of small boat sailing in the 1960s and '70s, the idea of special ensigns has become something of an anachronism. The custom undoubtedly pleases some, but the times for which the rules were made have changed and it would be difficult, if not impossible, to put the clock back.

It is not suggested that anything in particular needs to be done regarding the present use of special ensigns in UK waters although it is undesirable in principle to have rules that are frequently abused.

In view of the considerable confusion caused by the use of special ensigns by British yachts when overseas, because they are often thought to be Service craft or to have some other official role – that particular 'grey area' can easily be eliminated by encouraging all who are so entitled to use the Red Ensign as their national colours in foreign waters.

It is a flag that never causes confusion, and also one of which yachtsmen can be proud.

Where to wear an ensign?

Another of the many customs associated with the use of flags, and one to which the traditionalist will always want to cling, is that an ensign is worn, whereas any other flag is flown.

Nowadays an ensign is best worn on a staff at the stern. By custom it is also considered correct for a yawl or ketch to wear her ensign from the mizzen masthead, if there is no room to do so from an ensign staff, and a gaff-rigged yacht may use the peak of the gaff. However, the stern is most often the answer and the recent 'short cut' of having an ensign on the backstay, instead of on a staff, is unfortunate.

Most sea-going yachts have a stern pulpit of one sort or another and it is simple to incorporate an ensign staff socket into it.

House flags

Traditionally (and it is impossible to get away from tradition when writing about flags), the ensign showed a ship's nationality, but burgees and 'house' flags showed allegiance or actual ownership.

Some commercial vessels still fly house flags, which carry the company name or logo. However, the more recent custom of 'signalling' the company name with 'VIKING LINE', or whatever, in huge capital letters on the topsides amidships, has largely superseded the other tradition of having the company 'livery' in the form of a prominently painted funnel.

In a modern ship the exhaust fumes will be pumped out in some suitable manner – often via the mast/ventilator structure – but the huge striped funnel has gone the way of so much in ship design and it is highly unlikely ever to return.

Generally a house flag has little or no place on a yacht; it is a form of commercial identification. However, during a regatta some racing yachts will fly a very large, and usually brightly coloured, 'house' flag while in port or while motoring out to the racing area.

This rather pleasant and quite new custom (new in the sense that most flag etiquette goes back centuries) is akin to the plume on the helmet, or the design on the shield, of a tournament knight of the Middle Ages.

As a piece of sartorial bravado, and so long as it is not flown at sea, when it could be mistaken for a signal, the 'battle flag' is fun; especially when it is distinctive and brightly designed. The best ones are specially made to suit being flown from about halfway up the forestay and, for their very specialized role, there is no reason why they should not fill a sizeable part of the fore triangle.

If we ignore the little advertising pennants with 'Volvo-Penta' or 'Shell' written on them, which are an unfortunate trend in some European waters and have no place on a yacht whatever, one other type of 'house' flag is the comparatively recent custom of hoisting square flags with the RYA or the RNLI emblem on them. If a man feels the need to display to his fellows that (supposedly) he has joined his national authority, there is no reason whatever why he should not do so, but at sea that type of house flag could be easily mistaken for a signal flag and it most certainly should never be left flying when underway.

As for an RNLI 'house' flag, it would be difficult for anyone who goes to sea to justify not supporting the British lifeboat service, but I do suggest that the money spent on buying the flag would do more good in an RNLI collecting box.

As chairman of both an RYA and an RNLI Committee at the time of writing, I can hardly be accused of lack of interest in the affairs of either organization, but I cannot see that flying a little piece of bunting, day and night, until it disintegrates is any particular support to the organization concerned or has any sensible connection with the sometimes vitally important subject of signalling at sea.

Club burgees

In Britain each club has its own burgee and until very recently it would have been rare to see a yacht without a burgee flying from her masthead if there was anyone on board. The burgee is normally flown from a staff at the masthead in such a manner that the flag is free to rotate on the staff. The purpose is twofold. First, it tells others of the owner's allegiance: the club to which he belongs. When a man belongs to more than one club it would be normal to fly the burgee of the local club, when appropriate.

The second function is as a wind vane. Traditionally, the burgee is something to which the helmsman refers while sailing to help him to get the best out of the available wind. Recently, however, the introduction of electronic devices to display wind direction, the adoption of masthead-mounted navigation lights, and the introduction of VHF aerials which need to be placed at the masthead have all laid claims to this position.

Most cruising yachts still use burgees in UK waters – although the custom has never been so universally adopted in some other European countries.

Racing yachts do so less and less, and when they do, the helmsman is more likely to be watching the dials of his electronic vane rather than watching a flag.

The use of flags while racing is referred to in Chapter 13.

It goes without saying that a man cannot fly the burgee of a club of which he is not a member. The burgee is normally hoisted when the owner goes aboard and left flying while he is in effective control. The traditionalist would consider it bad manners to wear a special ensign (with a warrant from one club) and the burgee of another, and some would claim that the burgee should be raised and lowered with the ensign. But it is more logical to leave the burgee flying at night – if the owner is in effective control – in the same manner as the distinguishing flag of a flag officer in the Royal Navy is left flying day and night.

From the practical point of view some form of 'signal' at the masthead is a considerable help when handling the boat herself. From the racing point of view the electronic vane, which can have an amplified display close by the helmsman, and the strands of wool sewn into the luff of a sail may be the right answer, but most cruising helmsmen will look aloft in any case.

When manoeuvring in confined spaces in a marina it can be of very considerable assistance to be able to glance aloft and check what effect the wind is likely to be having on the movement of the boat. In a large lock, for example, when the wind on deck might have little or no effect, the wind in the rigging might make all the difference to the decision to go port-side-to or starboard-side-to the lock wall. A burgee at the masthead is the most foolproof way of getting that information.

Dressing ship

The custom of dressing overall during a special occasion must be one that goes back centuries, and although the traditionalist may enjoy quoting the precise order in which the flags of the International Code are *supposed* to be placed, it is more realistic to suggest that it no longer matters at all.

What does matter is that if celebrating HM The Queen's official birthday, the opening of the local regatta, or merely a party because a friend is launching a boat, the flags should be clean and in good condition and of an appropriate size for the vessel on which they are used. (see figure 10).

The RYA booklet *Flags and Signals* includes the traditional customs regarding the use of ensigns at the masthead as well as aft, and all sorts of other details, but the chance of finding that the precise length of the hoist, when all the code flags are strung together, actually fits the length of stay from stem to masthead to the stern is unlikely, to say the least.

The connoisseur will have a special set of code flags made up on its own line (with equal spacing between each flag) for use for ceremonial occasions. At a more practical level, the owner of a yacht about 10 m l.o.a. will find that he can hoist 12 to 14 mixed alphabetical and numeral flags between his bow pulpit and masthead, and 13 or 15 between masthead and stern pulpit. Those numbers are based on the assumption that he uses flags the Admiralty classifies as Size 6. It is unwise to carry anything much smaller than that – regardless of what they look like when held in the hand. In figure 10 the alphabetical flags are about 0.5 m in the hoist. The numerals are shorter in the hoist, of course, although very much longer overall, and the substitutes are slightly greater in the hoist than the letters, but all are about 0.8 m overall in height, clip to clip. A hoist for dressing ship looks best if the letters and numerals are alternated and the predominant colours mixed up a bit.

In my opinion this approach, which is perfectly correct for an

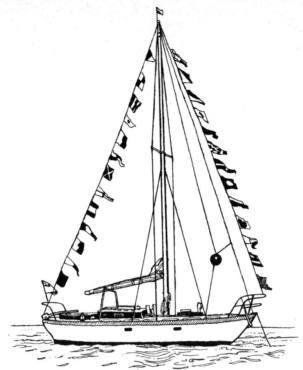

Figure 10 *When 'dressing overall' either for a National Festival or for a local occasion it is the way the flags are hoisted, not any particular order, that matters most. The space available must be full (but not overfilled) with sensible sized flags. These, on a 10 m l.o.a. hull are about 0.5 m in the hoist and the ensign is 0.6 m in the hoist; 1.2 m in length.*

informal occasion, applies equally well to any more formal one and, needless to say, the club burgee, a courtesy ensign if appropriate, and the ship's own ensign will be set at the same time.

Finally, note that dressing is an ancient custom – and rather a jolly one – but it is decoration. Never use dressing flags while underway.

When to use flags

From what has already been said, it must be clear that there are few hard and fast rules concerning flags. (Also, while writing on a subject so steeped in maritime tradition, it is of interest to note that even the phrase 'hard and fast' must be yet another of the dozens of English phrases with maritime origins.)

Generally burgees are flown when the owner or his party are on board and in effective control. House flags, on yachts, are 'special cases' and used in regattas.

Ensigns, on the other hand, do have a depth of tradition associated with them. They should be hoisted at 0800, when Daylight Saving Time (DST) is in force and at 0900 in the winter months. They are lowered at sunset or at 2100 local time, whichever is the earlier.

Overseas the customs are similar to our own, but it has to be admitted that the days when the times of 'colours' were strictly observed by the majority are long gone. It is pleasant, for those who attempt to maintain the 'colours' routine while cruising, to wave a brief 'Good morning' to the possibly equally sleepy occupant of the next-door boat, if he too has struggled out of a warm bunk to hoist his ensign promptly at 0800. However, it is the wearing of proper colours when overseas, and the wearing of courtesy ensigns, that matters far more than the precise timing of any flag-hoisting ceremony.

Special flags, on the other hand, can have special rules. The International Code Q, for example, which is required in certain circumstances when first entering foreign waters, and also required in all circumstances when returning to British territorial waters from overseas, must be left flying, regardless of the time of day, and the UK regulations require the flag to be 'suitably illuminated'.

Requirements of the law: Ensigns and Code Q

So far as the law is concerned, an ensign, to be used in accordance with the requirements of the Merchant Shipping Act and the Code Q, just mentioned, are the only two signals that are mandatory for all yachts.

Until recently all ships throughout the world used Code Q in the meaning it has in the International Code of Signals. Any vessel entering a country from another asked for clearance 'Pratique' by flying Q until she had been boarded and cleared by the appropriate health authorities.

Nowadays, such health checks as there are on commercial craft are arranged through shore-based agents and by radio, and the use of flag Q indicates a request for Customs and Excise clearance. All yachts, whether they are registered or not, and both foreign and British, must fly Code Q 'on entering United Kingdom territorial waters'.

If separate health clearance should be needed, the signal is Code QQ, which should be made by hoisting Q and the first substitute.

The procedure that follows the hoisting of Code Q varies with the circumstances (1982), and is somewhat outside the scope of this book in any case, but for UK citizens the next step may be to notify the local Customs and Excise office and that has to be done within two hours of arrival in the port. The justification for mentioning the procedure in a book about signalling is that this notification (which is a signal) can be by radiotelephone, once the yacht has arrived in port, by means of a link-call to the local office via a coast radio station. This facility can be a great convenience although no reference to the use of radio appears in the official document Customs and Excise Notice No. 8.

Overseas the requirements of the law affecting Customs and Excise signals vary from country to country.

In **France**, provided that the visit is legitimate and that no goods are carried that should be declared for customs purposes, yachts arriving from foreign waters on their own bottoms may enter the country without reporting to the authorities on arrival or on departure. If they do have dutiable stores to be embarked they *are* required to report, but in neither case is the Code Q required.

In **Belgium** Code Q must be flown when first entering the country from overseas, and the regulations state that the yacht will be boarded by Customs Officers. In practice, however, it is the Belgian immigration authorities that are more likely to board visiting yachts.

In the **Netherlands** Code Q need be flown only when bonded stores are on board, but the regulations state that on arrival a member of the crew should report to Customs.

In **West Germany** a yacht arriving from a Scandinavian or European Economic Community (EEC) country need not fly Code Q, but others should. An alternative is to sail direct for the Kiel Canal and fly third substitute, day and night, which is to indicate that Customs clearance has not been obtained. If stopping in the country, however, Customs must be cleared at one of a small number of main entry ports.

If in any doubt, Code Q is an international signal and anyone entering another country's waters is signalling the state of his passage by hoisting it. However, many countries require the first port of entry to be a main one, at which there is a customs office, and increasingly the onus is on the visitor to take the initiative and make contact. Thus, overseas as well as at home, a radiotelephone call might save much wandering around in the rain.

13 Signalling and yacht racing

Since the very beginning of the sport, yachtsmen have used the code flags of the day for starting their races and they have always given them private meanings of their own. If the custom was not so well established all over the world, it could be described as both foolish and dangerous. But the custom evolved because, up until the first world war at least, racing usually took place well away from any commercial shipping.

In the early growth days of offshore racing the Royal Ocean Racing Club had three classes and every yacht was required to race with either Code X, Y or Z in her starboard rigging. At the time those letters had different meanings in the International Code of Signals to what they have today, but they still had meanings to which seamen all over the world would be expected to react.

Some correspondence in *Yachting World* in the 1950s drew attention to the undesirability of the situation. It was also pointed out that as there were three classes and three substitute flags it would be simple to change the rule.

That was done and for many years the three classes used the three substitute flags for class identification until the size of the fleet called for more classes. There was then a natural, and logical, transfer to today's custom of flying numeral pennants to identify the classes.

In Chapter 12 it was pointed out that an ensign flown from a backstay is a somewhat unsatisfactory signal. It does not hang properly and looks what it is, a makeshift. Numeral pendants, on the other hand, are very much shorter in the hoist when compared with their length, and the modern custom of flying them from the backstay, about 2–3 m above the deck, has reason. The alternative – flying class identification flags from the crosstrees – has the overwhelming disadvantage that, when racing sails are used in the foretriangle, they will rarely be seen.

Possibility of confusion

On shore, as well as on an anchored yacht which is being used by a race officer, International code flags are regularly used as class flags and, although the custom is well established in small craft history, it undoubtedly can, and on at least one occasion already has, caused severe problems. In view of the increase in the amount of yacht racing, and remembering that many races are started in waters shared with commercial craft, it is important to consider the implications.

Even if the problem is minimal when code flags are displayed from a clubhouse or other shore establishment, it can only be undesirable for a yacht, anchored in, say, a commercial harbour and being used for the start of a race, to fly an International Code flag alone. The preparatory signal – P – may have an innocuous meaning, if any passing ship did wish to read the signal and to react, but many of the individual class flags have very important meanings. The International 505 Class, for example, has W as its class flag with *I require medical assistance* as its international meaning. That particular anomaly has already caused rescue services to be called out off Chichester Harbour with unfortunate recriminations.

The simplest way to avoid a repetition of such misunderstandings is to draw attention to the alternative provision within the existing rule where the warning signal is the class flag 'or distinctive signal'.

There are many ways of avoiding the possibility of confusion and subsequent accusations of irresponsibility. The best is to use numerals. Another would be to fly a private flag superior to the class flag – a club burgee for example – and thus make it obvious that the signal was a private one and was not using code flags in their normal sense.

It is accepted by custom that a yacht club is 'a registered British ship'. It wears the ensign of the club (the maritime flag) and never the Union Flag, as might any other building ashore. Therefore the club ensign should be hoisted and lowered as though it were on a ship.

International Yacht Racing Union

Yacht racing throughout the world is controlled by the International Yacht Racing Union and it is the IYRU rules that everyone uses. Individual nations sometimes add their own prescriptions to those rules, but generally this is only to clarify local customs – such as the use of class flags; the basic rules themselves rarely change.

From the signalling point of view Rule 3 states that any change in the sailing instructions can be made only in writing and not later than the warning signal for a boat's class has been given. Oral instructions are not accepted as a valid signal unless given in accordance with a procedure previously laid down in the sailing instructions.

IYRU rules 4 and 5 cover signalling procedure and they are reprinted in full in Appendix G. One important, and not always understood, point is that it is always the visual signals that govern. However, for the start of a really important event such as the Fastnet Race, the organizers write additional information into the Race Instructions and radio is used for guidance both for a count-down to the start and to assist with recalls.

Yacht race management

The expansion in dinghy racing during the 1950s led to huge fleets, as well as a racing season that extended far beyond the traditional three to four months in the summer. This in turn brought about the presence of the rescue/safety boat as a norm, rather than as an exception. As a guideline the RYA recommends one rescue boat for every 15 competitors in the small centreboard classes. The consequence has been the utilization of radio for yacht race management and, with the law in the United Kingdom concerning the use of radio changing as much as it has, the practice is certain to develop and expand.

Radio for inshore races: UK waters

Up until 1982 the most common legal way for rescue/safety boats to keep in touch with each other and with the race officer ashore was by the use of channel M – 157.85 MHz: the procedure was explained in Chapter 5. Ch. M was first allocated for mainly social use by marinas but subsequently licences were given to certain clubs to allow them to use the frequency for yacht race management.

In theory, if there was adequate discipline and only a limited number of clubs and marinas using the facility, the concession – what is really a private yachtsman's 'port operation' channel – would be excellent. Unfortunately, the system is often overtaxed by a mixture of ignorance of the need for discipline and blatant disregard for the needs of others. By no means is Ch. M a reliable communications tool.

One alternative has been the use of 27 MHz AM 'walkie-talkie' (usually called Citizens' Band) but not only is this illegal in many countries, including the UK, but it too is subject to a great deal of interference and a generally 'cowboy' attitude to discipline.

Towards the end of 1981 the UK authorities licensed the use of 27 MHz frequency-modulated (FM) equipment (see Appendix C) and this is likely to prove better than the illegal 27 MHz amplitude-modulated (AM) sets. However, the buying public did not respond as well as many had thought they would. Also there is already evidence that the 'cowboy' element will spoil the use of 27 MHz FM in much the same way as 27 MHz AM has been spoiled in the United States.

Yet another answer, for somewhat more specialized purposes, is to apply for a special frequency. The regulations allow for the issue of a Private Mobile Radio (PMR) licence. It would normally be issued only with good reason and it has the disadvantage – when compared with Ch. M – that it can be used only by, say, the flag officer's boat or the rescue boats that are listed in the licence. However, despite the additional cost (£65 in 1982), it has the tremendous advantage of being exclusive. The RYA has one such PMR licence for use for certain special events.

Next in the long list of alternative solutions, one club in south east England has licensed *all* its safety boats with Ship Licences. Thus they can communicate between themselves, or with other yachts, on international intership frequencies. One slight disadvantage of using maritime VHF frequencies for yacht race management is that all operators must have a Home Office certificate to operate whereas for Ch. M only there is not that restriction. And another minor difficulty is that a vessel using intership channels cannot also use them for communication with a race officer ashore. However, given common sense and good discipline, the idea of having a Ship Licence for each safety boat has much to commend it.

Finally, yet another solution for the management of an inshore yacht race may well prove to be the use of the second group of frequencies that were licensed by the Home Office in 1981. In addition to the 40 channels now licensed in the 27 MHz band for FM equipment, there are 20 channels in the 934 MHz band. Suitable equipment in that band was not available for the 1982 season but, in the longer term, there is considerable technical evidence to suggest that the 934 MHz sets may solve many of the race officer's problems.

(See Appendix C for a summary of the use of Citizens' Band frequencies and the ranges that might be expected.)

Radio for offshore races: UK waters

As has already been mentioned, Ch. M has been used by the Royal Ocean Racing Club for the start of the Fastnet Race and in the past the club has employed 2301 kHz as well for broadcasting from a committee boat. However, the new single-sideband regulations, which came into force in January 1982, now make that impossible.

Other changes for the 1979 and 1981 Fastnet Races included the use of mother ships and a scheduled twice-daily reporting system on VHF. The mother ships were equipped with medium frequency R/T as well as VHF R/T and they were thus able to relay the state of the race and/or to report any anxieties to the organizers ashore.

That particular schedule applied only to the competitors in the Admiral's Cup, although in 1981 all competitors were required to report when rounding the Fastnet Rock. Thus 1981 marked the first time that all yachts had been required to report in while racing in UK waters.

Following the 1979 Fastnet Race storm and the clamour (much of it ill-informed) concerning the need for radio, the new RORC regulations went a small way towards the provision of safety cover for the fleet. However, it should be pointed out that the media interest in that particular event and the constant enquiries regarding the positions of the competitors also had a part to play.

Radio for offshore racing: Overseas

In some parts of the world, notably Australia, radio reporting in races has been in existence for many years, but the reasons are complex.

One factor is that during the leading yacht races in Australia there is a tremendous public interest in the progress of the event, some of it associated with the amount of gambling on the result. Another reason is that, because of the greater distances involved, Australia has not been able to develop the continuous VHF coastal cover found in Britain. This, coupled with considerably less stringent type-approval standards for MF radio, has encouraged the use of MF equipment with its greatly increased range.

With MF radio available comparatively cheaply, and no VHF R/T on the coast, the idea of using mother ships and regular reporting schedules on MF bands has therefore grown up with the development of the sport.

Radio watchkeeping in the future

For events like the occasional round-the-world races, which usually involve heavily sponsored and thus very publicity-conscious skippers, there is bound to be a great interest in the sophisticated equipment which will allow a man to keep in touch with the shore from the cabin of a comparatively small yacht over many hundreds, and sometimes thousands of miles. However, although the yachts in some of the events may not be so very large compared with commercial vessels, their radio equipment is comparable.

For the 'ordinary' offshore event, on the other hand, even the relaxation in the type-approval standards for MF radio in the United Kingdom (see Appendix D) which was one of the spin-off benefits that followed the furore over the 1979 Fastnet Race (and which allows UK yachts to install MF R/T, at about half the cost of the equipment that was type-approved previously) is still not the answer.

MF R/T, which has a range of say, 150 to 200 miles, has obvious advantages, both from the reporting-in and from the overall safety points of view, but it is still expensive to buy and comparatively complicated to install. VHF R/T, on the other hand, is not expensive, nor is it difficult to install. We now have a state of affairs where practically every yacht racing offshore has VHF R/T fitted, but neither the organizers of offshore events, nor the competitors, seem to have realized the potential advantages of using what they have.

Reporting-in

The RORC in 1982 began introducing the idea of a reporting-in schedule on a voluntary basis. But although reporting-in has obvious advantages in reassuring organizers or relatives who might otherwise be anxious, a reporting-in programme based on a twice-a-day schedule has little to commend it from the safety point of view.

If a competitor is out of radio range of the shore but within range of many of his competitors, and he wants to ask for assistance, there is little to reassure him in the realization that not one of his fellow yachtsmen is likely to be listening until the next reporting-in schedule. Even then, without very strict discipline, the air will be full of traffic from navigators anxiously plotting each other's positions.

Compulsory reporting-in implies mother ships, which are by no means always available, and it also introduces the unanswerable problem of what to do if a yacht fails to report. There are numerous reasons why a yacht might not be able to report at a particular time, and a decision about whether or not to try to send assistance that was based on the *lack* of a report is an upside-down way of reacting in any case.

Listening watch: Racing

A racing fleet, where all vessels are equipped with radio, ought to be able to keep in touch with itself; to look after its own from the safety point of view.

There is no silence period on VHF as there is, worldwide, on MF R/T, but any system of listening watch would work only if the race organization introduced one.

The object of a silence period is that everyone is requested to listen, and to refrain from transmitting, for three minutes. For the professional on MF the times are three minutes after every hour and every half hour. I am suggesting that for a start there should be a voluntary silence period, among a racing fleet, for three minutes after 0033, 0633, 1403 and 1803.

Those times are chosen (1982) to follow after the main shipping forecasts (when the skipper or navigator is likely to be using his receiving radio in any case) and, for the benefit of those who might be listening to them, to avoid the official silence periods on MF R/T at every hour and half hour.

The objective is to build in periods when any skipper who was anxious or who needed help had a reasonable chance of being heard.

Whether the listening watch would be on Ch. 16 or on the intership frequency Ch. 72 (which the RORC hopes will become accepted 'by common use' as the intership channel for racing yachts offshore) is a subject for discussion, but the principle is clear.

The next step along the same road would be the introduction of cockpit-mounted loudspeakers as the norm, because the cost of installing them is comparatively negligible. Then the schedule of listening watch times might easily be varied or extended without any disturbance to those on their watch below.

If the four-times-a-day schedule was acceptable – and for some reason the amateur sailor seems to resist the idea of keeping a radio watch although the professional today takes it as a matter of course – the next step might be for an offshore race organizer to ask that if a shipping forecast had included winds of perhaps Force 6 or more, then Schedule B was to operate; competitors would then be asked to listen, say, every two hours.

The point is made. In distress a yacht can call on others by numerous means, but there are many degrees of assistance or reassurance that a yacht skipper might need to call upon.

The idea of using VHF R/T as a means of telephoning home, or keeping in touch with the office, seems well established in the minds of most yachtsmen, but the idea of extending the facility they already have to provide a safety 'net' for the offshore fleet of which they are a part has still to be developed.

Chapter 8 referred to the principle of keeping a listening watch while cruising, but when racing the priorities are bound to be somewhat different. It is not suggested that the three-minute listening watch, four times a day, would be the ideal, but it would be a start.

14 Navigation lights and shapes

The purpose of a navigation light, or a group of navigation lights is to signal: *Presence, Aspect,* and *Occupation.*

By day a shape, or a group of shapes signal: *Presence* and *Occupation.*

Presence

The prime requirement for a vessel is to be seen. Most of the small craft navigation lights on chandlers' shelves have 10 W bulbs in the white lights and 25 W in the coloured. Today's lights were developed during the early 1970s and, although not many yachtsmen seem aware of the fact, since July 1981 it has been illegal to use navigation lights that have not been type-approved by the Government testing station. Some yachts still have the earlier types fitted but the four-year period of grace, after the new regulations became law, has expired.

Because the coloured screens in a navigation light filter out about 80 per cent of the available light (see Chapter 9) it is hardly surprising that small craft lights in the past had such a bad reputation. Rarely were they fitted with bulbs larger than 5 W, which is approximately 5 candela. Thus with 80 per cent of the light lost there was theoretically 1 candela outside the lens; and that only if the lens was clean and the battery well topped up.

Voltage drop is somewhat outside the scope of this book – it is primarily an electrical installation problem – but the result of allowing a battery to get too low is so catastrophic it is important to understand what happens.

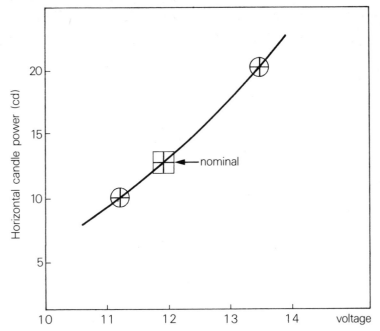

Figure 11 With a 10w bulb and a nominal 12 volt battery the bulb will give off about 12.5 cd. In fact if the battery is chock full or the alternator is running, the output is about 20 cd compared with a mere 10 cd if the battery is less than half full and the engine off.

The drop in the horizontal candle power show in figure 11 is for a 10 W bulb. The nominal 12-volt battery actually has about 13.5 volts when on charge, or when completely full, but the voltage soon drops and is only about 11 volts when the battery is low. If there is then a further drop caused by using electric cable of too small a cross-section, it is hardly surprising that the ability of a small craft sidelight to show 'presence' was sometimes queried.

Aspect

The aspect of a ship is shown at night because all navigation lights are designed around the same simple geometry (figure 12). The Steering and Sailing sections of the Collision Regulations are well beyond the scope of this book and it is not my intention to try to summarize them.

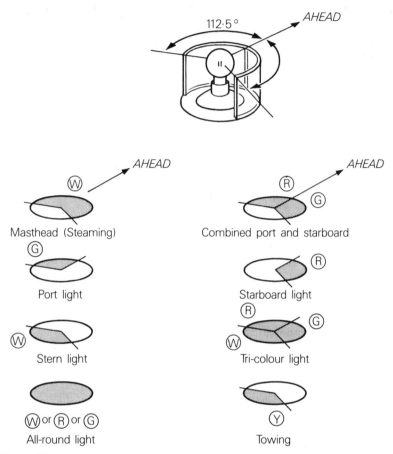

Figure 12

However, it is a fundamental principle of the rules that a vessel is considered to be an overtaking vessel (and thus liable to give way) if she approaches from more than 22.5 degrees abaft the beam. Thus, if the approaching vessel can see the sidelights of another then she is crossing, but if she can see only the stern light she is the overtaking vessel.

Note that the 22.5 degrees abaft the beam applies both to the sailing yacht's sidelights and the sidelights of the largest tanker. The brilliance (and thus the range) will vary but not the angles.

Figures 13 and 14 show the general layout of navigation lights, as well

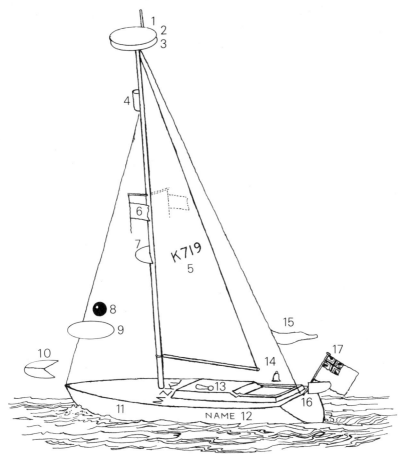

Figure 13 Auxiliary sailing yacht

1. VHF aerial
2. tri-coloured light
3. all-round white (signalling)
4. radar reflector
5. national letter and sail number
6. Code and courtesy flags
7. masthead (steaming)
8. anchor ball
9. anchor light
10. combined port and starboard
11. sail number or call sign on deck
12. name on topsides
13. foghorn
14. anchor bell (portable)
15. racing class flag
16. stern light
17. ensign

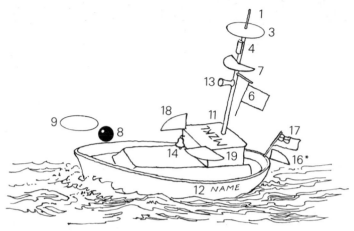

Figure 14 Motor yacht *Lit in a similar manner to the arrangement shown in Fig. 13 (p. 135) except that the motor yacht will not install a tri-colour light, she will not exhibit a sail number, nor will she have a racing class flag. Also motor yachts usually fit separate port and starboard lights. Numbers refer to same devices as in the sailing yacht with the addition of 18. Starboard light and 19. Port – but it is perfectly legal for vessels of less than 20 m to use a combined port and starboard if they prefer.*

Note:
From 1 June 1983 vessels of less than 12 m l.o.a. may, when under power, combine the masthead (steaming) light (7) and the sternlight (16) into one all-round light at the masthead (3) to be used with sidelights. Thus the auxiliary sailing yacht will be allowed to motor with (3) and (10) and she need not fit (7) and (16). The motor yacht, similarly, can use (3), (18) and (19) and she, too, need not fit (7) and (16).

This change is a considerable help because it takes the sternlight away from the low mounting aft – so often on a transom where it is partially masked by wash or exhaust smoke – and puts it where it can be seen.

The disadvantage of the simplified lay-out is that all the navigation light 'white eggs' are in one basket.

Figure 15

Although a vessel under oars is required merely to show a torch a powered dinghy should fit a mast (the transom is probably the best mounting) and exhibit an all-round white light.

as other signals, on small motor and sailing boats, and figure 15 applies to dinghies.

Note that the auxiliary sailing yacht has a slightly more complicated installation than the motor yacht because, being an auxiliary, she has two contrasting occupations to show: sail and power. Note, too, that the simplified requirements for dinghies under oars and for motor dinghies of less than 7 knots show merely presence; they display neither aspect nor occupation.

Occupation

Of the three requirements, occupation is probably the most complicated because the decisions as to who *gives way* and who *stands on* is frequently governed either by the vessel's occupation or by the state she is in at the time.

At its most simple a sailing yacht's 'occupation' by day is shown by her sails. If she is motoring as well, she is required by the rules to indicate that she is also 'propelled by machinery' by exhibiting a black cone (Rule 25 (e)). Then, although the implication is not always appreciated by yachtsmen, she is a power-driven vessel as far as the regulations are concerned. (See Motor sailing below.)

One other 'grey area' in the rules concerns vessels being towed. It seems illogical to require a vessel being towed to exhibit 'sidelights and a sternlight' – Rule 24(e) – when they are also the lights required for a vessel under sail. However, in the re-writing of that particular rule for the current Collision Regulations, the existence of the yellow towing light – Rule 24(a)(iv) – does go a small way towards the need to differentiate between a vessel or object being towed and a vessel under sail. Both are normally 'stand-on' vessels but the anomaly is there nevertheless.

Occupations that restrict

Figure 16 shows a few examples of the more specialized signals used to define vessels that have varying degrees of priority because of their occupation. A vessel that is restricted in her ability to manoeuvre *because of the nature of her work* exhibits red, white, red all-round lights – or diamond, ball, diamond by day – in addition to her normal lights.

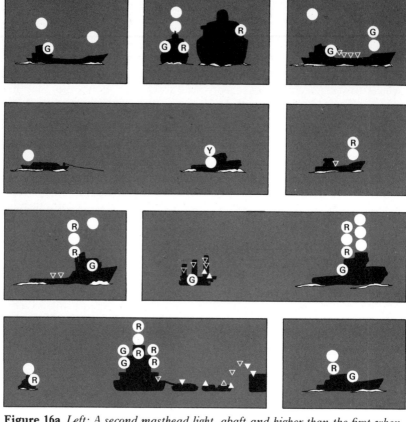

Figure 16a *Left:* A second masthead light, abaft and higher than the first when above 50 m. *Centre:* When towing the vessel exhibits two, or if the tow exceeds 200 m, three masthead lights. *Right:* When trawling a vessel exhibits all-round green over white and, if over 50 m a masthead light. She is also likely to be showing deck working lights.

Figure 16b When towing the vessel will exhibit a yellow towing light vertically above the normal sternlight. Where practicable the tow also exhibits side lights and a sternlight. *Right:* Fishing, other than trawling, all-round red over white.

Figure 16c A vessel making way but restricted in her ability to manoeuvre by the nature of her work exhibits all-round red, white, red, in addition to her normal navigation lights. She, too, would probably be showing working deck lights on the after part of the vessel. *Right:* A vessel towing unable to deviate from her course exhibits the all-round red, white, red in addition to her normal towing lights; tow more than 200 m.

Figure 16d A vessel dredging is also restricted in her ability to manoeuvre and exhibits the all-round red, white, red and also exhibits two red lights to indicate an obstruction and two green on the side on which it is safe to pass. *Right:* A pilot vessel on duty and underway.

That signal might be seen on a vessel laying a navigation buoy, for example. The next degree of priority is the vessel *restricted in her ability to manoeuvre such as renders her unable to deviate from the course.* She, too, exhibits the red, white, red but her usual navigation lights would normally include the three masthead lights in a vertical line. This signal is for tugs with exceptional tows, such as oil rigs.

The third in this group is the vessel *engaged in dredging.* She shows the same red, white, red, but she adds two red lights, one above the other, on the side on which there is obstruction, and two green lights on the side where vessels may pass. By day the dredger has the diamond, ball, diamond to replace the red, white, red lights and the obstructed and safe sides are marked by two diamonds for the obstructed side and two balls for the safe.

Yet another type of restricted ship is the vessel *constrained by her draught.* She exhibits a black cylinder by day and three all-round red lights in a vertical line by night.

Finally in this summary of signals showing occupation, a vessel that is *not under command* exhibits two black balls by day, two red all-round lights by night and, when making way, she also exhibits sidelights and a sternlight.

Note, however, that although not under command (n.u.c.) may be a signal which many small craft skippers will never see in a lifetime of sea-going, the other three signals using red, white, red (to indicate a restriction in the ability to manoeuvre) and the three red lights (for a vessel with deep draught) are comparatively common. As far as small craft themselves are concerned, there is an escape clause, agreed in 1982, stating that vessels of less than 12 m l.o.a. need not exhibit the extra n.u.c. or vessel aground lights and shapes, nor need vessels not normally engaged in towing show the extra masthead lights or the yellow towing light *if towing because of distress or a need for assistance.*

Working lights

From the legal point of view the Collision Regulations are the result of a great deal of careful thought and, with the possible exception of Rule 25(e), are clear. Some would like to see an extension of the power to exhibit red, white, red, when towing, to give them greater freedom of action, but the 1982 interpretations added to the 1972 rules make it perfectly clear that the signals are to be used only for major towing operations, and the tow of, say, an oil rig is likely to be so smothered in working lights that she is highly conspicuous in any case.

On the other hand, and from the small boat point of view, the working lights on fishing vessels as well as on many anchor-handling barges and other oil-rig supply boats are often highly confusing because of their brilliance. The fact that the bright deck lights are usually shining aft may eliminate most of the glare problem for the officer of the watch on the vessels themselves, but it must be assumed that they do not always realize how difficult it is for others, close astern of them, to see. Some cross-channel ferries also use powerful floodlights facing aft and it is to be hoped that the practice can be discouraged. It adds little or nothing to the safety of the vessel exhibiting the floodlights, but greatly affects the night vision of others, *near sea level*, who are relying on night vision for much of their pilotage.

Shore lights, too, can be a considerable hazard for the man seeing them from low down. Obviously it is efficient to illuminate a dock well, but harbour authorities need to take into consideration the glare from seaward and see that it is kept to a minimum.

Small craft lights and shapes

One problem which must be obvious to even the most casual observer is that there is no point in learning merely the signals the small craft herself may have to exhibit. There is a great deal else that has to be understood.

Chapter 9 had something to say about brilliance, size and the sound levels of signals. Chapter 10 referred to sound signals and how they are used at sea. This chapter, so far, has referred to navigation lights and shapes in general. Rule 3, Part C and Part D of the Collision Regulations with Annex I, II and III, shown in Appendix M, tell the story, but figure 17 summarizes the more important navigation light configurations. It illustrates the way that a vessel's 'occupation' is shown by her lights. The question of right of way is outside the scope of this book – and Part B of the Collision Regulations applies – but even if small craft will normally be expected to keep clear of specialized vessels, such as those engaged in towing or in fishing, they have many rights, too. The small craft's rights, however, can be enjoyed only when she, too, shows her presence, aspect and occupation in an unambiguous manner. Figure

Authorized and customary visual signalling methods.

Vessel at anchor – One ball, forward

Vessel at anchor – All-round light

'I require assistance' – Code V

Motor-sailing, as power-driven vessel

'Request Customs clearance' – Code Q

Sailing vessel underway – Tri-colour

Morse signalling lamp

Foreign courtesy and National ensign

Figure 17a

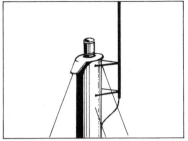
Radio telephone – VHF dipole aerial

Attracting attention – White flare

Sailing vessel, restricted visibility

Distress – Orange smoke

Motor-sailing – by day, one cone

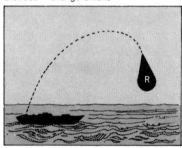
Parachute distress rocket

Distress – Square flag and ball

Distress acknowledged – 3 white at 1 min

Figure 17b

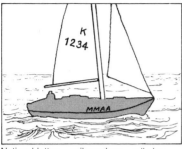
National letter: sail number: call sign

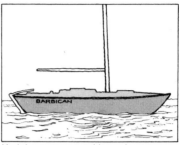
Yacht's name on topsides or house

Distress – Hand-held flare

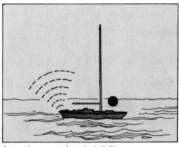
At anchor, restricted visibility

Figure 17c

13 for the auxiliary sailing yacht and figure 14 for the motor yacht show 'model' layouts.

To those whose initial reaction might be: 'But what on earth do I need all that for?' the answer is simple. 'You don't; at any one time.' Figures 13 and 14 represent all the signals a well-equipped yacht might ever need to make.

As another way of looking at the same problem, figure 17 is a selection of all the essential, useful and not-so-useful signals and signalling methods that are included in the Collision Regulations and the 'ordinary practice of seamen'. Note, however that figure 17 includes the distress signals, which are the subject of the next chapter.

One interesting point affecting figures 13 and 14 is that from June 1983 the regulations are modified to allow small craft to combine the existing masthead (steaming) light and the sternlight into one all-round light at the masthead. Both the advantages and disadvantages of the change are explained in the caption.

Observant readers may have noticed that whenever the masthead light – white and shining through 225 degrees – has been mentioned, the word 'steaming' has always been added. That is because what is now called the masthead light was called the steaming light in earlier editions of the rules and the expression is still a common part of everyday speech. It is unfortunate that this potential confusion should exist because we now have the masthead light, the tri-colour light – mounted at the masthead and often referred to as 'the masthead light' by yachtsmen – and we also have the all-round white light just referred to which will also be mounted at the masthead.

The only simple solution would be to continue to use the term 'masthead light' for the 225-degree light, to call the tri-colour light by that name, and to refer to the 'all-round white' by that name. Both the tri-colour and the all-round white can be bought in a combined fitting with the 10 W bulb in the white part and the 25 W for the tri-coloured.

Congestion at the masthead

The change in the regulations which will allow a masthead (steaming) light and a stern light to be combined into one all-round white light, (which has just been referred to) will encourage those who have not already done so for signalling purposes, to install an all-round white light at the masthead. However, with wind direction and speed indicators needing masthead units, with VHF aerials at their best if mounted at the masthead, with tri-coloured and all-round white lights having to be at the masthead and with the antenna for position fixing devices (such as Decca and Loran) all requiring as unobstructed a site as possible clear above any mast, the congestion at the top is a major problem.

For signalling, for wind information and for navigation the efficiency of the installation is obviously important but it is not something that is covered by law. For navigation lights, on the other hand, not only are the International Regulations for Preventing Collisions at Sea affected, the safety of the vessel is as well. If a navigation light is obscured it might as well not be there and a light that is intermittently obscured is almost as bad, because it can confuse as well as be a danger.

It is Appendix I 9(b) which applies. It states that all-round lights "shall be so located as not to be obscured by masts, topmasts or structures within angular sectors of more than 6 degrees, . . ."

In practice a tri-colour and an all-round white light can be bought in one fitting (one on top of the other) and one, three-core cable is enough because the two bulbs share a common earth wire.

A VHF aerial need not be thicker than 2 to 3 mm if the flexible stainless steel dipoles aerials are used but even the more common 'broomstick' dipole is only about 25 mm in dia. Thus a stand-off bracket of 15 to 20 cm from the side wall of the mast will mount the aerial far enough away. Of the two, the new type of whip with a small encapsulated matching unit has many advantages both for weight, wind resistance and visibility. However the lack of a stand-off bracket inhibits the hoisting of a burgee.

The wind direction and speed masthead unit can be a problem but most are now made with a very thin arm to carry the actual unit up and forward of the masthead. The point to watch is that a thin forward-sloping arm must be positioned so that the obstruction to the all-round light or lights is within acceptable limits. Sometimes, however, the unit itself is quite bulky and the all-round light and the mounting bracket must be so arranged to avoid having the mass of the actual wind unit causing unacceptable arcs of obscurity.

Finally beware of antenna such as the type being produced for some of the latest position fixing devices. The Decca Yacht Navigator antenna for example – which the makers recommend should be mounted clear above the mast – is 50 mm in diameter.

Motor sailing

It is not the intention of this book to discuss the implications of the manoeuvring side of regulations – merely the signals associated with them. However, it has to be admitted that Rule 25(e) – the rule that requires a black cone, exhibited forward, when a sailing boat is also being 'propelled by machinery' – is not a rule with which it is easy to comply. Unless the foresails are lowered and the black cone hoisted in the foretriangle, it will not generally be seen from many aspects. A sensible interpretation of the rule – but not one that would necessarily stand up in court – is that if a sailing vessel is being propelled by her *sails* as the principal form of propulsion, it is reasonable that she should *not* exhibit the cone. However, when the *engine* becomes the principal form of propulsion then she should; otherwise she is asking for a 'priority' from others to which she is not entitled.

In UK waters the rule requiring a black cone is frequently ignored, but it must be pointed out that it remains the legal requirement and in the event of a collision the lack of one would undoubtedly be taken into account by a court or by an insurance company. It is also important to note that in some neighbouring countries, especially Germany, the rules for cones are strictly enforced by the maritime police.

Anchoring lights and shapes

Earlier in this chapter – under the subheading 'Occupation' – it was pointed out that one of the most basic 'signals' of all is that of a vessel under sail. It is the sails that show she is a sailing vessel. Similarly, a power-driven vessel is assumed to be what she is unless she exhibits something to change her category, and the most basic change is the anchor ball. A vessel at anchor is no longer underway, and thus has priorities in the sense that others take action to give way to her.

For reasons that I fail to understand, many yachtsmen ignore the requirement to show that they are at anchor, by day, even if few are so foolish as to do so by night. The rule concerned is perfectly reasonable. If away from a fairway, narrow channel or anchorage and away from 'where other vessels normally navigate', the anchor ball need not be used. Otherwise it is required, and to anchor without one is somewhat akin to walking along a country lane at dusk with a dark grey mackintosh and the vague assumption that passing cars will be able to see you in time to take the necessary avoiding action.

Figure 18 indicates that an anchor ball does not have to be the 0.6 m dia. required for larger vessels, but may be 'commensurate with the size of the vessel'. One convenient answer is an inflatable 'ball' which is about 0.4 m in dia. and readily available on the UK market and there are also several types of collapsible balls. Even an old pullover pushed into a string shopping bag (or a dark fender) would be a lot better than nothing!

To prevent the ball, or the anchor light for that matter, from banging about and chafing itself against the forestay, figure 18 also shows how triangulation with a snap shackle on the forestay and a downhaul keeps all quiet.

Figure 18 *Vessels of less than 20 m l.o.a. may use shapes "commensurate with the size of the vessel". Here, on a yacht about 10 m l.o.a. a black anchor ball 0.4 m in diameter 'looks-about-right'. For comparison the code flag/courtesy flag in the crosstrees is about 0.5 m in the hoist and the ensign 'looks-about-right' at 0.6 m in the hoist; 1.2 m in length.*

Note: *The anchor ball is triangulated, to prevent it 'banging about': the same can be done for a paraffin anchor light.*

Nowadays, when battery capacity is not quite such a difficulty as it used to be, many yachtsmen use electric anchor lights, but on a sailing yacht, where continual battery charging sessions can be an irksome chore, an oil anchor light can be perfectly satisfactory if it is of a type made for the job. An ironmonger's so-called hurricane lamp is *not* suitable because it is likely to soot up or blow out in strong winds if not adjusted carefully, but a properly made anchor light with internal cone and a type of double-walled chimney seems to stay alight whatever the weather.

Beware of the small 'yachtie' sizes, however; they soot up too easily and the paraffin containers are not really large enough. A wick of about 15 mm should be the absolute minimum and one of 20 mm wide is much better. That size means a lantern which is about 0.4 m high, but a large lantern with the wick adjusted not too high is far more satisfactory, and more likely to remain constant, than a small one adjusted to its limit.

One other device associated with anchor signals which is well worth its house-room is a photo-electric switch built into an electric anchor light. With a photo-electric switch a vessel can be left safely when the crew are ashore in the knowledge that the anchor light will be switched on at the suitable time. The saving in battery drain in the early mornings is also an advantage because the light switches itself off when the crew are still undisturbed in their bunks.

Summary of sizes

With a realistic allowance for batteries that are not full and coloured screens that are a trifle salt-spray encrusted, a white light with a 10 W bulb will be visible about 2.5 nautical miles and a 25 W bulb in a coloured light, and similar allowances, will be visible about 2 nautical miles. It is unwise to expect anything better.

For shapes, an anchor ball that is, say, 0.4 m in diameter looks about right.

As a general rule auxiliary yachts need to carry most of the numerous special signals summarized by figure 17. Code flags for occasional use for signalling or for dressing overall should not be less than 0.5 m in the hoist. A courtesy ensign of about 0.4 to 0.5 m also seems sensible, although some chandlers offer 'miniatures' that I find a bit silly.

An ensign for a 9–10 m cruising yacht looks about right if it is 0.6 m in the hoist and 1.2 m long. Finally, for the single-letter visual signals a yacht might make (and distress is dealt with in the next chapter) a code flag such as Q (for Customs and Excise clearance) or V (I require assistance) can be as big as 1 m in the hoist without being too large. (See also figure 25.)

Many yachtsmen try to signal with flags appreciably smaller than the dimensions given, but that may explain why it is so common to hear complaints of signals that were never seen.

Yet another reason why small craft signals are often not seen is that all

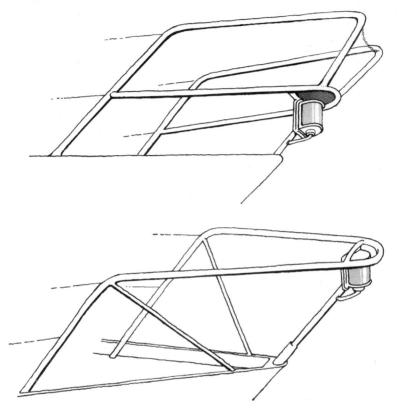

Figure 19 *Two examples of navigation lights fitted in sensible positions. They will not be obscured and are reasonably well protected.*

too frequently they are mounted where either the sails or part of the yacht herself masks the lantern. It is useless, for example, to mount navigation lights on the side of the coachroof or deckhouse. It was one of the traditional places but, when the yacht is hard on the wind, the lee light will be masked by the genoa and the windward one by the hull itself.

For a sailing yacht the masthead and the bow pulpit are the only satisfactory places (see figures 13 and 19). That is why the new rule, 23(c)(i), allowing the masthead (steaming) light and the stern light to be combined into one all-round light at the masthead (and used with a combined port and starboard light in the bow pulpit) is so welcome.

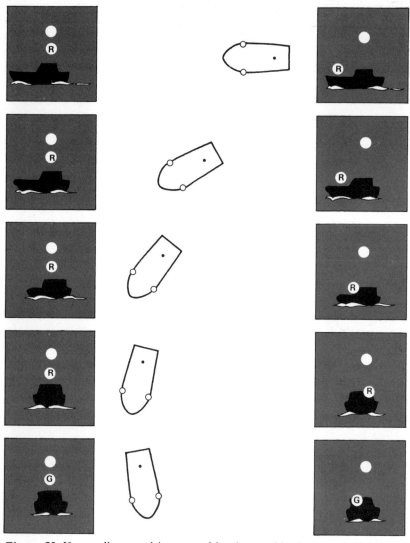

Figure 20 *If a small power-driven vessel has her combined port and starboard light mounted vertically below her masthead light she gives no indication of change of course.*

For a small motor yacht (or motor-sailer which normally uses her engine when on passage) the new regulation will be even more welcome because, prior to its introduction, so many small motor yachts fitted their stern lights either on the transom, where they were far too low, or on the after side of the deckhouse where they often interfered with the helmsman's night vision.

Note, however, that although the masthead is the ideal place for the new all-round white light, the mast is *not* the best place for a combined port and starboard (figure 20).

PART 3 Emergency and other special topics

Chapter 15: Distress and urgency 153
More than one type of signalling method — How many? — When to fire pyrotechnics — Search by aircraft — Other visual signals — Urgency signals — Signals to attract attention — EPIRBs — 2182 kHz EPIRB — Reflective and retro-reflective signals — Where to stow distress and urgency signals — The need to report the use of distress and urgency signals.

Chapter 16: Radar reflectors, Life-saving, Port operation and Night vision 171
Life-saving signals — Whistle — Man-overboard lights — Port traffic signals — Gale and strong wind warnings — Blue light — Medical transport — The practicalities of signals and signalling — Night vision.

15 Distress and urgency

If her radio is still operational and if the vessel in distress is within radio range of the station to which she wishes to speak, a radio is obviously the first choice of distress 'tool'. However, as a means of attracting attention when in distress a red parachute rocket has several advantages; not the least is that it can alert ships or shore stations that are in the vicinity but who might not have been listening to the radio.

A parachute distress rocket ejects a brilliant red flare of about 30 000 candela to a height of about 300 m and, because the rocket is attached to a miniature parachute, it hangs in the air for about 40 seconds. Of all the internationally accepted signals the distress rocket is probably the most instantly recognized for what it is.

As with any distress signal, a distress rocket signifies that: *A ship is threatened by grave and imminent danger and requests immediate assistance*. It should never be used as an urgency signal, but many of the problems of our rescue services – both professional and volunteers like the Royal National Lifeboat Institution – arise because there is confusion. An urgency signal signifies that: *The calling station has a very urgent message to transmit concerning the safety of a ship . . . or the safety of a person.*

To take an extreme example in circumstances such as were found during the 1979 Fastnet Race storm: because some yachts, although obviously in need of assistance, had no means of signalling the state they were in, valuable helicopter rescue time was lost by pilots having to winch men down to find out what was required. If a rescue helicopter is engaged in that manner unnecessarily, it is not available to go to the

153

assistance of another crew nearby where the need may well be far greater.

More than one type of signalling method

It is not the intention to repeat what was said in Part 1 concerning distress and urgency procedures on VHF. Chapter 4 referred to the technique concerning distress calls and the priorities. Chapter 6 was devoted to distress and urgency procedure. And Chapter 8 described the co-ordinating role of HM Coastguard.

However, that all concerned radio only but it has to be admitted that in a small vessel, in distress, it is the radio itself which is likely to be one of the first casualties. The aerial is vulnerable, the set itself is susceptible to moisture, and the lead acid battery might either be affected by bilge water or merely have become flat through over-use. Therefore, even if a small craft in distress has been able to alert the rescue authorities by radio, it is by no means certain that she will be able to continue to remain in contact.

Furthermore, despite the emphasis that was placed in Chapter 5 on the importance of transmitting the position, it is highly unlikely that that position will be correct. Evidence suggests that a few moments before a boat actually goes aground the navigator was almost certainly not aware that he was in that position. If he had been aware he would usually have been able to do something about it. It is therefore reasonable to assume that a position given from a casualty is suspect to say the least.

A sophisticated answer to this dilemma is for the rescue authorities to take D/F bearings of the signal when it is originally received, and equipment to allow bearings to be taken on VHF transmissions is being installed on the more congested parts of our coastline. The French also have a similar facility. However, this is a new development (1982) and it most certainly does not apply to all distress traffic yet.

Having transmitted the initial alert, the most likely signals as a back-up are the distress rocket and the distress flare. As has already been mentioned, the rocket goes high and therefore it is in itself an alerting device. The hand-held flare, on the other hand, is sometimes described as a 'precise location' signal. Although it is also an alerting device, for those close by who happen to be looking in the right direction, it does not burn so brightly as a rocket, nor does it have the same impact on the eye.

Both signals are useful but both have disadvantages. The rocket can disappear into low cloud, so that most if not all its alerting capability is lost. The hand-held flare, on the other hand, will have a very restricted range if it has to be used at water level. There is no perfect answer.

How many?

Yet another problem – apart from the ever-present one of cost — is deciding on the number of flares that are to be carried. Even with a strictly limited number of signals available, it is recommended that they should normally be used in pairs. The argument for so doing is based on the fact that a distress signal may be noticed for only a few seconds before it goes out. It will come as a complete surprise. 'Good heavens,' the mind will say, 'was that a red flare?' and it is reasonable to assume that the observer will then scan the horizon in the general direction of where he thought he saw the signal. A second signal – say 30 seconds after the first has extinguished itself – will confirm the first glimpse and allow an approximate bearing to be taken.

Distress at sea is not something that can be practised. Rescue procedures can be rehearsed and the rescue services go through many hours of training, but the ordinary man may never see a distress rocket fired in earnest in ten years of cruising.

I am sure the 'fire twice' philosophy is sound. Despite experience of sailing going back to childhood, the first time I saw a rocket fired 'in anger' was on a clear dark night when it was perfectly obvious what was seen. However, three of us in the cockpit – including one professional ship's officer and one Dutch maritime police officer – reacted with amazement. After it had died down I went below to the radio to report what I had seen and the others turned the yacht around, but none of us took a bearing. The incident occurred during a race and we had rounded the south end of the Kentish Knock sands about half an hour before. Thus, a rocket somewhere on our starboard quarter was guessed, correctly as it turned out, to have come from a yacht on the Kentish Knock, but the lesson is one I hope I never forget.

The Royal Ocean Racing Club and the RYA make similar recommendations regarding the minimum number of flares to be carried. For coastal cruising the RYA minimum is two rockets, two hand flares and two orange smoke signals. For offshore cruising the RYA and the RORC recommended minimum is four rockets, four hand flares and two smoke signals.

Twice as many would not be excessive.

When to fire pyrotechnics

Obviously the first pair of rockets would be sent off as soon as the skipper had satisfied himself that he really was in distress; but, assuming that there were only four or six rockets aboard, it would probably be counter-productive to fire off the lot. After two the advice would be to wait, keep a good look-out, and try to sort yourself out.

When a distress signal has been seen by the shore authorities it will be answered, by night, by three white rockets emitting white stars, sent at about one-minute intervals. By no means will that answering signal always be seen but, where there is a reasonable chance that it will be spotted, it will be made . The most important first step to take, after the receipt of a distress signal, is to send an acknowledgement.

If no acknowledgement of the initial distress signals has been seen, and no other ships are in the vicinity, the decision regarding further signals is difficult, but the available evidence suggests that most people fire off their signals too quickly rather than too slowly. The time interval before a second pair of rockets is fired would have to depend on the circumstances – and this book is concerned with signals rather than with seamanship – but where possible it is best to keep the precise-location flares until a ship is seen to be in the vicinity, or until you have reason to believe a ship is actually searching for you.

For an example: a lifeboat may send up a white parachute illuminating rocket at intervals or, if the radio was still operational, the Coastguard would probably ask the casualty to send off a flare when the lifeboat was thought to be in her vicinity.

Search by aircraft

In any major incident, such as the search that followed the 1979 Fastnet Race storm, a fixed-wing aircraft may be used to establish the position of a vessel or group of vessels in distress. The aircraft will fly a search pattern (see figure 21) and will fire green flares at each turning mark, and periodically during a leg of her course at 5- to 10-minute intervals.

Search and Rescue helicopters are also used for this task and they too follow the same principle although, obviously, the length of a search leg, the gaps between any two legs, and the height of the aircraft will all depend on the circumstances, the visibility and the type of aircraft.

These green flares should be 'answered' by the vessel in distress. The official advice is to wait until the glare of the green flare has died down (to help the crew recover their night vision); fire one red flare; fire another red flare after about 20 seconds (to allow the crew to line up on your bearing); fire a third red flare when the aircraft is overhead or appears to be going badly off course. From this advice it is obvious that a stock of four precise-location hand flares is unquestionably a minimum.

An alternative recommendation for those who have no choice, in the sense that they have only one or two flares left, is to wait until about 30 seconds after the glow of the green flare has died down and then fire one red flare, but, having understood the principle of an aircraft search pattern, the circumstances will have to control what action is taken.

As a general rule civilian aircraft are unlikely to be involved in SAR activities, although they would report any signals they had seen or heard – and in any case civilian aircraft would not be equipped with maritime radiotelephone frequencies. SAR aircraft, on the other hand, do

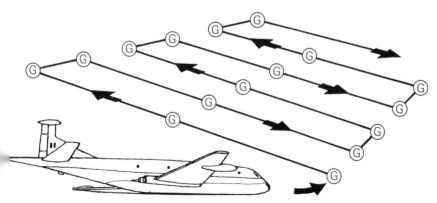

Figure 21 *If an aircraft is searching it may fly a pattern, firing green flares at intervals along each leg (the number depending on height) and at each turning point. If you see a green flare, wait about 30 seconds after it has gone out, for the crew to recover their night vision, then answer with a red flare.*

generally have maritime VHF sets and in British waters an RAF Nimrod involved in a major incident, or any of the Navy or RAF helicopters would normally have Ch. 16 and 06.

Figure 3 on page 75 showed the complexity of the frequencies used during a search or rescue incident, and because an SAR helicopter has VHF Ch. 16, do not simply assume she is listening on that frequency. It is HM Coastguard who is likely to be co-ordinating a search and indiscriminate calls to overflying helicopters asking 'What is going on?' will not be popular. In the first place the pilot of the helicopter is more likely to be using Ch. 00 if he is working with, say, a lifeboat and HMCG, during a search. Alternatively he may be working on other frequencies with his own base.

When a pilot is actually engaged in a rescue, Ch. 16 is likely to be used as it allows the helicopter to speak directly to the casualty, to the lifeboat and to any other ships involved. However, it is most important that the radio should never be used when the helicopter is winching. At that moment the helicopter pilot cannot see the casualty beneath him and, while actually winching he is being fed with information by his own winchman via his intercom to allow him to hover in the precise position required.

Never transmit on the frequency that a helicopter is using during winching.

Other visual signals

So far, in this chapter, only radio or pyrotechnics have been mentioned but, as figure 22 shows, there are a large number of internationally recognized distress signals. From the five marked * from the list, radio and pyrotechnics are the two most important but the signal consisting of 'a square flag having above or below it a ball or anything resembling a ball' has enormous advantages over all other visual distance signals because although pyrotechnics have the alerting capability, they do not go on giving their message. A distance signal like the flag and ball can be recognized for what it is even if the crew are lying unconscious down below.

In practice an anchor ball and any flag is all that is required.

Figure 23 illustrates the only satisfactory way in which a man on an

Figure 22: *Annex IV* International Regulations for Preventing Collisions at Sea. Distress Signals

1. The following signals, used or exhibited either together or separately, indicate distress and need of assistance:
    - **(a)** A gun or other explosive signal fired at intervals of about a minute.
    - **(b)** A continuous sounding with any fog-signalling apparatus.
    - **(c)** Rockets or shells, throwing red stars fired one at a time at short intervals.
    - **(d)** A signal made by radiotelegraphy or by any other signalling method consisting of the group . . . _ _ _ . . . (SOS) in the Morse Code.
    - ***(e)** A signal sent by radiotelephony consisting of the spoken word 'Mayday'.
    - **(f)** The International Code Signal of distress indicated by N.C.
    - ***(g)** A signal consisting of a square flag having above or below it a ball or anything resembling a ball.
    - **(h)** Flames on the vessel (as from a burning tar barrel, oil barrel, etc.).
    - ***(i)** A rocket parachute flare or a hand flare showing a red light.
    - ***(j)** A smoke signal giving off orange-coloured smoke.
    - ***(k)** Slowly and repeatedly raising and lowering arms outstretched to each side.
    - **(l)** The radiotelegraph alarm signal.
    - **(m)** The radiotelephone alarm signal.
    - **(n)** Signals transmitted by emergency position-indicating radio beacons.

2. The use or exhibition of any of the foregoing signals except for the purpose of indicating distress and need of assistance and the use of other signals which may be confused with any of the above signals is prohibited.

3. Attention is drawn to the relevant sections of the International Code of Signals, the Merchant Ship Search and Rescue Manual and the following signals:
    - (*a*) A piece of orange-coloured canvas with either a black square and circle or other appropriate symbol (for identification from the air).
    - (*b*) A dye marker.

upturned dinghy or a sailboard could indicate distress. The signal – repeatedly raising and lowering the arms – was added to the international list for the 1965 edition of the Collision Regulations and it needs to be given wide publicity because it serves so well for the fastest-growing section of the sport: board sailing. The emphasis must be on the phrase 'slowly and repeatedly'. The signal is not a 'hello' wave between friends. That is normally made with one arm only, and for just a few seconds.

In the same category are the pair of signals shown in figure 24 which are used by sub-aqua swimmers. For the 'I am OK' signal the arm is held still – as it is for the traditional greeting between the officers of the watch on passing ships. There it is common to see a man on the wing of a

*Signals likely to be of most interest to small craft.

Slowly and repeatedly raising and lowering arms outstretched to each side.

Figure 23 BODY SIGNALS: BOARD SAILING
International distress signal applicable to dinghy or board sailing.

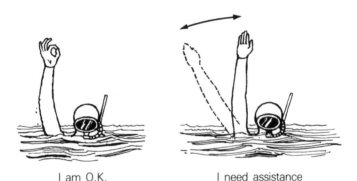

I am O.K.   I need assistance

Figure 24 SKIN DIVING *(Sub Aqua)*
The skin diver may make the "I am O.K." signal to craft approaching too closely, merely to emphasize his presence.

A *diver's boat will exhibit Code flag "A", or a rigid replica, not less than 1 m in the hoist.*

bridge raise one arm and hold it aloft for two or three seconds and it is answered in a similar fashion.

When assistance is required, the sub-aqua swimmer will move his arm rhythmically sideways.

Urgency signals

As was briefly explained at the beginning of this chapter, there is an important difference between distress and urgency. The master of a vessel is legally obliged to go to the assistance of others in distress, and amateur sailors, too, are included in that precept. For anyone to use a distress signal when not 'in grave and imminent danger' simply means that any facility sent to his assistance might be removed from another vessel which could be in very much greater need of quick action. Once the, usually slender, resources have been committed there may be nothing else available.

In addition, and apart from those short-term practicalities, many of the rescue services in northwest European waters are completely voluntary, and most of what is not voluntary is enormously expensive. If the facilities are used as a 'get-you-home' service – however tempting it might be to persuade oneself that the situation is worse than it is – the whole structure of the service is put at risk. People do not readily volunteer if others indulge in 'crying wolf'.

The most effective urgency signal for yachtsmen (other than the use of radio as described in Chapter 6) would be the use of a large code flag 'V' by day, or the flashing of Morse V by night. Figures 25(a) and (b)

Figure 25a *A single-letter code flag 1 m in the hoist looks enormous, when held in the hand, but it looks what it is – about right – when hoisted aloft.*

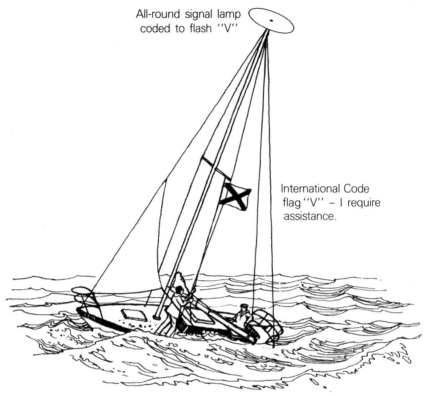

Figure 25b *If things start to go wrong and assistance is required there is not much chance of having a signal read if the code flag is less than 1 m in the hoist. Here a 10 m l.o.a. yacht is flying "V" an urgency signal comparable to a radio PAN-PAN call. At night an all-round signal light could be coded to flash "V", continuously.*

illustrate the signal. The idea of having a simple device that can be wired into the circuit for a masthead-mounted all-round white light, and is coded to flash V, has been discussed with one of the leading navigation light-fitting manufacturers and is likely to be available in 1983. The great advantage of the idea is that, like the distance distress signal, an urgency signal at the masthead would continue to send out its message even when the crew are otherwise engaged.

The fact that an urgency signal, like a distress signal, is something which will only very rarely be required does not detract from the need to understand its meaning. It represents a concern that would not justify the use of a distress signal because the yacht was not in 'grave and imminent danger', but where the yacht was in urgent need of assistance.

To a certain extent it can be misleading to describe imaginary problems and then offer solutions, but if a motor yacht had a rope round her propeller and the skipper had sprained his arm trying to sort himself out, that yacht might well be in urgent need of assistance. However, unless she was close to windward of an exposed and rocky shore, or there was some other special circumstance, it would not be right to consider her as being in distress. In that imaginary circumstance an urgency signal might well be the correct answer.

A passing vessel could then either investigate herself, or pass a message to the shore, or both. If the casualty had radio she would be in a position to explain the nature of the urgency, but assuming that a lifeboat was to be sent, a signal – by day or night – identifying the casualty as the vessel which was in need of assistance, would be likely to save time and trouble for all.

The flashing of code V – *I require assistance* – from the masthead with an automatic coding device would satisfy that need admirably.

Signals to attract attention

It is reasonable to suppose that, since the beginning of time, the seaman has used a flare-up light of one sort or another to attract attention. *Flames on the vessel* . . . are to this day an international distress signal, but flares are also used to attract attention and up until the present edition of the Collision Regulations the rule referring to the attracting of attention – now Rule 36 – has mentioned 'a flare-up light'.

In practical terms, this has meant the occasional use of the hand-held white flare as a means of drawing attention to the presence of a vessel to another which had not seen her.

During the 25 years prior to the introduction of the present type of small-craft navigation light, when, because of the growth of the sport, small boat sailing at night was becoming commonplace, white flares were also looked upon as a normal part of this activity. To this day, the Royal Ocean Racing Club requires all yachts to carry them, and it is right that the club should do so. In my own experience of night sailing in the 1950s and 1960s I would estimate that I might have used an average of one white flare per night. In contrast, since sailing with the modern tri-colour light at the masthead – fitted with a 25 W bulb compared with the 5 W maximum that used to be the norm – I have only once found the need to use a white hand flare. Such is the difference that good lights have made.

In the 1972 version of Rule 36 the wording is different though the meaning is the same: 'If necessary to attract the attention of another vessel any vessel may make light or sound signals that cannot be mistaken for any signal authorized elsewhere in these rules, or may direct the beam of her searchlight in the direction of the danger, in such a way as not to embarrass any vessel.'

However, in the United States and in Australia a practice has developed among some yachtsmen of using masthead-mounted strobe lights as a means of attracting attention. The strobe (xenon) light is intensely bright and it certainly attracts attention, but the light was being used and left flashing in shipping lanes in a manner that other mariners found highly confusing.

One of the reasons for the use of the strobe was that the US yachtsmen did not adopt the new generation of navigation light in the way that they were accepted in European waters. Thus, while still using the 'tired glow-worms', he did not really expect to be seen. Furthermore, the Cardinal buoyage system had not then been adopted in either US or Australian waters and so the idea of fitting a bright flashing light (which flashed at exactly the same rate as a North Cardinal buoy) did not have the impact that Europeans might have expected.

During the late 1970s the subject was discussed at length in several IMCO committees and despite the objections from the US delegation, the recommendation, from all but the Americans, was that Rule 36 should have added to the original text: 'Any light to attract the attention of another vessel shall be such that it cannot be mistaken for any navigation aid to navigation. For the purpose of this rule the use of high intensity intermittent or revolving lights such as strobe lights, shall be avoided.'

By June 1983 this becomes international law. American advertising in 1982 was still referring to the imagined advantage of using strobe lights in traffic lanes for 'collision avoidance'. How it can be thought to be desirable for a yacht to turn herself into a North Cardinal buoy is not explained!

IMO has acted as described, and it is to be hoped that those who have been advocating the use of strobe lights for 'collision avoidance' will realize the problems they would create.

It is the navigation lights themselves that show presence; the flare-up, or similar, signals are an additional and preferably very occasional signal.

Flashing lights should, as far as possible, be reserved for navigation marks.

EPIRBs

Yet another emergency device that can be of interest to yachtsmen is the Emergency Position Indicating Radio Beacon (EPIRB) but, as with many new devices, there are serious misconceptions concerning its use. EPIRBs, or Personal Locator Beacons (PLB) as they are occasionally called, were developed to allow the rescue services to locate the position of a crew after an aircraft has crashed in the sea. For this purpose they are highly efficient.

The great difference in the circumstances between an aircraft casualty and a marine casualty is that if the former is in trouble that fact is likely to be known to her shore base within minutes. Even in the unlikely event that the aircraft is not able to alert her base regarding her difficulties, the fact that she did not report in on schedule will be known very quickly.

The small boat, on the other hand, is on her own. She might be at sea, or away from her normal base for several days, and no one would raise an eyebrow.

The key point concerning those diametrically opposed circumstances is that an EPIRB has no alerting capability. It is designed to locate, not to alert.

In addition it must be remembered that an EPIRB transmits on the aeronautical distress frequencies 121.5 MHz (for civilian aircraft) and 243 MHz (for military), and although an overflying plane would probably pick up an EPIRB signal, all civilian aircraft switch to Air Traffic Control frequencies as they approach the European continent. Within 150 to 200 nautical miles of our shores, no *civilian* aircraft would be listening.

It is true that there is a slight chance that a military aircraft might be within range, and certain coastal airfields might be alerted by near-by signals, but the chances are very slim. It is completely wrong therefore to consider an EPIRB to be anything other than what it is: a position-indicating device for use *after* an alert has been broadcast.

One exception to this state of affairs, worth bearing in mind for those crossing oceans, is that overflying aircraft do monitor 121.5 MHz when well away from land. Whether the authorities would be in a position to do anything about a report or not is another matter, but a signal from a yacht in distress well away from land would probably be reported, although civilian aircraft could give only a very approximate position.

At present the whole subject of EPIRBs is under discussion. In the not too distant future it is probable that satellite-controlled position fixing techniques will fix the position of a distress signal and relay it to the shore authorities with coded information as to the type of vessel in distress. At this moment, HMCG does not monitor aeronautical distress frequencies, nor does the RNLI have equipment which can 'home' on them, but RNLI evaluation trials (1982) are being carried out on equipment to give the new Fast Afloat boats a homing capability on 121.5 MHz in addition to Ch. 06 and 16.

In Scandinavia and in parts of the North American continent the rescue services do monitor the two frequencies affected. However, for UK waters, an EPIRB is of interest for yachtsmen planning really long passages and as a back-up. It remains, nevertheless, highly irresponsible to think of an EPIRB working on aeronautical frequencies as an alerting device for small craft.

The authorities are so concerned about the problems and false alarms that could be caused by the misuse of aeronautical frequencies that a special M Notice (M 982) was issued in 1981. It gave advice and warnings concerning the use of the equipment – the carriage of which needs a licence – and in particular the importance of careful stowage. An EPIRB will transmit its signal continuously, once it has been set off, until either it is switched off or the batteries run down. Careless stowage can allow a set to be switched on accidentally with the all too obvious problems that would cause.

In home waters, on the other hand, if an alert had been transmitted by other means and an SAR helicopter was involved in a rescue attempt, it would undoubtedly increase the chance of being found quickly if the helicopter could home on your beacon.

2182 kHz EPIRB

When referring to EPIRBs it is usual to mean the small devices just described which transmit on the aeronautical frequencies. Some have a voice capability as well as an automatic 'bleep', and some merely the auto signal which transmits for 30 to 50 seconds followed by silence from 30 to 60 seconds. However, there is another type of EPIRB which uses the maritime distress frequency 2182 kHz.

When 2182 kHz-only distress equipment first came on the market in the 1960s the present generation of VHF equipment was still on the drawing board. Thus 2182-only, especially if it had the facility to trigger the auto-alarm equipment (used by merchant vessels and all but the smallest fishing vessels), looked most promising.

The disadvantage of 2182-only equipment is that the range is poor because the power output from built-in batteries is so low; and, although there is a speech facility, and RNLI Lifeboats and HMCG monitor 2182 kHz, the present generation of portable equipment is little if any better in range than VHF R/T and appreciably bulkier to carry.

Reflective and retro-reflective signals

The sun is by no means always available for use as a signal, but, when it is, a heliograph is a highly effective means of drawing attention. The device – a highly polished mirror with guidelines marked on it – is held in such a manner that the sun's rays are reflected back towards the ship or aircraft.

The other type of reflecting device that is of considerable use is the retro-reflective material which reflects the rays of a searchlight back to the sender. It is composed of minute glass balls encapsulated in clear resin and is similar to the material used on most of the otherwise unlit road signs.

The multi-national company 3-M is one of the main suppliers and its material Scotchlite, comes in many forms. Even a small piece of a few square centimetres has a remarkable effect; it is highly desirable to have patches sewn or stuck to the shoulders and hood of an outer garment, as well as to life raft canopies and similar equipment, because even the comparatively weak light from a torch will be reflected back in an almost unbelievable manner when within a hundred metres or so. White Scotchlite reflects 180 times more than a plain white painted surface.

During 1983 the International Association of Lighthouse Authorities were discussing a code to allow retro-reflective materials to be used to identify un-lit buoys. Green and red patches will be used for the Lateral System starboard and port hand marks, and the black and yellow colours on Cardinal System marks will be represented by *blue* and yellow retro-reflective patches.

There is an obvious danger, if searchlights are used without thought, because a carelessly used searchlight can so easily ruin another man's night vision. However, Scotchlite is so effective that even a hand torch can be used to advantage to identify an unlit mark, when at close quarters.

Where to stow distress and urgency signals

Official recommendations, as well as advice such as is found in RORC Special Regulations, usually refer to storing distress flares in 'a waterproof container' and the manufacturers offer 'offshore packs' and the like with, supposedly, a full stock of pyrotechnics in a robust screw-topped plastics container.

To me that seems completely the wrong approach, and probably a relic of the days when flares were little better than cardboard tubes which had to be physically protected as well as kept well away from moisture. Nowadays distress flares are manufactured in a very efficient manner. They are composed of a.b.s. plastics tubes with neoprene 'O' rings as seals at top and bottom.

If a man wants a flare he wants *one*, not a boxful. If he has to unscrew the top of a box – in the dark and wet and in a hurry – he wants to know what it is that his hand is touching when he puts it in the box. So with one hand holding the box, another holding the lid, what is he supposed to do next? If he gets rid of the lid there is a probability that he cannot find it again to screw back on and he has to put a hand into the top of a box, without being able to see what he is touching. One minute later he will probably want a second rocket and the whole cycle has to be repeated. It is a nightmare.

Flares are far better stowed individually; each with a pair of non-corroding clips. The deckhead is the ideal place. It is clear of practically all moisture, and distress rockets and flares can be grouped and clipped between the deck beams where they are well away from physical damage.

If rockets are placed to one side and hand flares to another there is no possibility of confusion. One, and one only, can be available within seconds of the decision to use one.

A box is the last place to carry the white flares that have already been referred to in this chapter. A decision to use one, for example when it has suddenly become obvious that an approaching ship is not going to give way, or is altering course in a manner that suggests that she is not keeping a proper look-out, is likely to be almost instantaneous. The situation is not one where the skipper says 'I think we may need a white flare in about half a minute's time'. It is far more likley to be the result of the sudden realization that the other vessel has not seen you.

When a white flare is needed to attract attention it is somewhat akin to the use of a horn on a motorcar: the driver does not want to have to look for the horn button. Similarly, the wise skipper keeps a white flare, or two, within almost instant reach of the helmsman's position. Clips are the perfect answer. Some flares have magnetic components and so should be kept well clear of any compass.

Urgent flags, buoyant smoke signals and other items not so amenable to the use of clips can usually be stowed by using rubber shock cord and suitable eyes but one of the few disadvantages of the modern mass-produced small boat, with glass-reinforced plastics throughout, is that it is by no means as easy to screw in a lacing eye as it used to be in the days when everything on a small boat superstructure was wood.

Stowage of signalling equipment really is important. Canvas pockets can sometimes be the answer and 'Velcro' is occasionally another, but the objective should be for all emergency equipment to be located where the crew can find it without delay *and in the dark*. As soon as it is necessary to use a torch to find the torch, your 'system' has let you down.

Need to report distress and urgency visual signals

As a tailpiece to this discussion of distress and urgency, it is important to try to ensure that any use of distress signals should be reported to the appropriate authorities; in the UK that means HM Coastguard. A few reports of signals which originate from the shore are malicious, but that is a problem for the police rather than for the mariner. However, there is also a steady trickle of reliable reports of distress flares used at sea (which are unlikely to be malicious), but which are never traced to any vessels in distress despite, on occasion, many hours of lifeboat search time.

The explanation is probably that an inexperienced man gets into difficulties of one sort or another – possibly hitting the sands – and he sends off signals feeling himself to be in serious difficulties. Shortly afterwards he gybes clear, recovers the use of his engine (or whatever it was that caused him to act as he did) and sails on. To be charitable he may genuinely feel that, because he saw no immediate reaction, no one had observed his distress signal. In reality a large number of signals are seen but are never traced, and at least some must be from small craft where the skipper did not have the sense, or the moral courage, to try to ensure that the authorities knew the true situation.

Obviously, without radio it would be difficult to 'cancel' the distress signal until after most of the damage was done, but it is still desirable that the true position should be made clear; even hours later. Also, because white flares used to attract attention also get reported as distress flares, it would make sense if craft with VHF reported the use of white flares – if they had been used close to the land. The difficulty of so doing is minute compared with the waste of time and energy caused by lifeboat searches for boats not in distress.

16 Radar reflectors, Life-saving, Port operation and Night vision

Everything in this book so far, except for the brief mentions of the use of a heliograph from a life raft or to attaching retro-reflective materials to a raft canopy or to outer clothing, has concerned signals that convey a message of one sort or another. However, the radar reflector is really very similar to the patch of retro-reflective tape. It is passive but its job is to pick up a signal from something else – another ship's scanner – and reflect it back to the source of that signal.

Radar itself is outside my brief, but radar reflectors are such an essential part of small boat signalling equipment that they ought to be carried, in certain conditions, on even the smallest craft that goes offshore. Even the open fishing boat can show her presence by hoisting a radar reflector on a short mast stepped through a thwart, if she so wishes, but the greatest difficulty regarding radar reflectors for small craft is that the neat little fittings so often seen in chandler's shops are almost useless.

The efficiency of a reflector depends on its design and on its size. The strength of the signal varies as the fourth power of the reflector diameter. Thus, quite apart from the question of the type and the loss of signal when heeled, the doubling of the size increases the reflected signal by 16 times!

If an octahedral reflector is mounted point-up, it has 'thrown away' a third of its efficiency even when the vessel is upright.

Unfortunately the only answer to heeling that is available is to use the type of reflector that incorporates a cluster of corners. However, a unit incorporating a large number of small corner reflectors is substantially inferior to a cluster of the same overall dimensions made up of a smaller number of larger corners.

For a reflector on a major navigation buoy which is to be seen above the wave clutter at, say, 10 miles, the recommended minimum diameter is 1 m. At a more realistic level, a yacht should never use an octahedral 'cube' that is less than about 0.45 m across the diagonal.

The recommended minimum size for a small boat reflector is the equivalent of the 10 m^2 sphere and the minimum height is about 4 m; more is better. Reflectors kept rigged aloft are the best and with these minima, and if the seas are not running too high, the echo should be visible at about 3 miles.

Life-saving signals

In one sense most of Chapter 6 concerned itself with life saving, because distress and urgency signals affect the saving of life. There are also life-saving signals for use by those on shore and figure 26 is a summary of one procedure. Basically the same signals – meaning 'Yes' or 'No' – can be used to assist a small boat trying to land or to direct a rescue operation using either a breeches buoy or a line between a casualty and those trying to assist.

Whistle

Another simple signal, and one that is associated with man overboard, is the mouth-blown whistle. At night, or in fog, a whistle signal would be far easier to hear, and probably more easily identified as far as direction is concerned, than a shout. Whistles are usually carried on the shoulder of a lifejacket and it would be a wise precaution to fit one, if it is not there already, as well as to keep one sewn into a pocket at the neck or on the lapel of an oilskin suit.

Of all the items of equipment mentioned throughout this book, the pocket whistle is probably the cheapest and, in certain special circumstances, the most effective.

Man-overboard lights

In contrast to the whistle – and it would be difficult to find one that cost

Life saving signals

Vertical movement of white flag or of a white light or flare by night, or of both arms. Or the firing of a green star.

For guidance of small boats: Code letter 'K'

MEANING: Affirmative – line is held – made fast – haul away.

For guidance of small boats: This the best place to land.

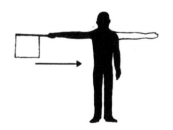
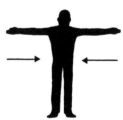

Horizontal movement of a white flag of a white light or flare by night or of both arms. Or the firing of a red star.

For the guidance of small boats: Code letter 'S'

MEANING: Negative – slack away – avast (stop) heaving.

For the guidance of small boats: Landing here is highly dangerous.

Figure 26

more than a comparatively nominal amount – man-overboard lights can be very expensive. At their simplest, a chemical-filled polythene tube (the light stick) has a thin glass tube within it so that when the outer tube is bent the inner breaks and the resultant mixtures of chemicals causes a dim green light. At their most expensive, man-overboard lights can be strobe flashing devices giving off intensely bright lights. Between these extremes there are many fixed and flashing models from which to choose, but they vary in reliability and must be checked frequently. For once, in the complicated and expensive business of equipping a boat for use offshore, the best is one of the cheapest.

Trials carried out by the RAF station at Plymouth in the late 1970s tested four different signals displayed by a man in the water in a moderate sea. Observers were stationed at different distances away to check the various reactions. Somewhat to everyone's surprise the two fixed lights of the four were found to be vastly better – the reason being that a light is visible only when it is on the crest of a wave. If the light is not lit at that precise moment it is useless.

The four signals tested were the comparatively cheap fixed torch bulb mounted on a floating container; the more expensive flashing version of the same device; the chemical 'light stick'; and the very expensive flashing strobe light. The clear order of priority was (1) the fixed 2.2 W bulb; (2) the chemical light stick; (3) the slow-flashing 3.5 W bulb device and (4) the strobe light.

From the air, needless to say, the strobe light was very effective, but from near the water level even the light stick, costing well under £1, was preferred to either flashing light.

Normally a floating man-overboard light is attached to any buoy. The RAF tests suggest that it should have a fixed light and that it need not be particularly bright because, for this situation, a great range is not the problem.

In view of the efficiency and the low cost of the chemical light sticks they, too, are an attractive proposition and one sewn into a pocket on the breast of an oilskin might make all the difference between finding a man overboard or not. However, the problem is that they are ignited by bending the tube and therefore, if put in a pocket unprotected, they would almost certainly be set off while working on deck.

One answer would be to store the light stick in a very light alloy tube. Otherwise there is too big a risk that the light stick would become bent, and thus useless, without the wearer realizing what had happened until he came to try to use it.

Port traffic signals

From the small boat navigator's point of view the majority of the small creeks and harbours used by yachtsmen have no port signals and no need for them. Nevertheless, with commercial ships being built larger and larger, many commercial ports have recently introduced signals. Also some of the man-made harbours with narrow entrances have had some form of traffic signalling since their beginnings.

What is surprising is that, despite a conference as far back as 1936, which was supposed to agree on a standard, there has been very little standardization around the maritime world.

Up until 1983 the Belgian and French harbours used one standard, which employed shapes by day and coloured lights by night, but there the problem was that although the signals were the same in several neighbouring ports, the way those signals were interpreted for use by small vessels was by no means the same.

Until now it has been necessary for the mariner to look up the meanings of traffic signals in his 'Pilots' or almanacs and also to try to find out if the signals also applied to small craft. In most ports they do; occasionally they do not.

Before long, it is to be hoped, all this uncertainty will be gone. The International Association of Lighthouse Authorities (IALA) has made proposals for port traffic signals. They are based on the use of light only, and because of the flexibility that has been built into the system, it is perfectly straightforward for a harbour authority to make it clear which signals apply to which class of vessel. These are at appendix N.

At present the ports that have signal shapes use balls or cones of about 1 m diameter. However, as was explained in Chapter 9, it is necessary for the mariner to be within, say, a mile to read that type of signal. One of the many advantages of the new proposals is that, where necessary, it will be possible to install lights which even in daylight can be seen very much further off than that. The first of the new IALA port traffic signals were installed at Dover and at Shoreham early in 1983. Several French and Belgian Channel ports are also expected to have changed by the end of 1983.

Tidal height is also signalled on occasion with a pattern of shapes or lights from some large commercial ports but the development of harbour radio will probably restrict any further spread of that particular custom.

For details of the existing harbour traffic or tidal height signals the appropriate 'Pilot' must be consulted but strong wind warnings, where they exist, are normally uniform throughout a country.

Gale and strong wind warnings

In the UK a black cone, point-up or point-down, is still used in certain locations to show winds of gale force expected from any point north or south of an east–west line. In a very few places the signal is replaced by a triangle of three white lights by night.

However, although our neighbours also have strong wind signals they are not uniform and not used by all ports. In the Netherlands, for example, weather signals are now displayed only at the Hoek van Holland and at Vlissingen (Flushing) and the Dutch use lights both by day and night. The Belgian ports exhibit two black cones points together or a blue flashing light when the wind from seaward is Force 4 or more, and when small craft under about 5 m l.o.a. may not proceed to sea.

Blue light

Flashing blue lights are found on police boats, fire tenders and on many lifeboats throughout northwest Europe. Generally it can be assumed that a blue flashing light will not be used unless the vessel is involved in some emergency work.

Medical transport

By international agreement the familiar Red Cross, and the alternative Red Crescent (used by many Muslim countries), is restricted for use by medical transport.

The practicalities of signals and signalling

When it is considered that the numerous methods of making signals which we have today have evolved over such a long period it is remarkable, in many ways, that there is so little ambiguity or conflict. The maritime distress system is remarkably efficient and small craft are as much a part of the overall maritime scene as the largest tanker in the world. Nevertheless, the problems on board the bulk carrier and the small yacht are very different and, if his equipment is to remain operational, the small-boat skipper has to be ever vigilant.

Perhaps the most obvious potential problem is that when he wants his radio the most – when dismasted or when he has serious leaks – the radio is at least likely to be serviceable. As a basic minimum the small craft should carry an emergency aerial – to be plugged in in case of spar failure – and an emergency source of power is also a wise precaution. Alternatively a second set, run from internal batteries, will give emergency cover, even if the range is limited. There is also emergency equipment on the market that normally runs from the ship's batteries, but from the internal supply for a limited period if necessary.

Thought must be given, too, to the installation of all signalling equipment. VHF can usually be mounted on the deckhead, where it is least likely to be affected by moisture, and an emergency portable would be best stowed in a sealed polythene bag.

The stowage of pyrotechnics was referred to in Chapter 15. They have the advantage that they will continue to work if all electrical circuits have failed, but electrical problems are often a problem because of careless installation rather than real disaster.

The batteries must be protected, and an 'umbrella' type cover is usually quite simple to arrange. Remember that a battery needs ventilation, but that need not prevent the cover shedding any drips. Make sure that the battery is not stowed too low in the bilge and either fit twin banks of batteries – charged separately by means of a blocking diode – or have an alternative means of charging if the auxiliary cannot be started by hand.

The latter solution is quite an attractive one because there are several suitable charging units on the market. The small ones are petrol driven – so there is probably the need to carry another fuel – but they are light as well as small, and the knowledge that there is the means to charge the battery, if it is every allowed to become flat, is a blessing.

The most sophisticated electrical equipment in the world is obviously little more than ballast if there is no electricity.

Remember, too, if all else fails and it is necessary to fall back on the use of distress rockets, that they do not behave 'logically' in strong winds. A rocket climbs *into* the wind as it goes up. Thus it should be pointed 10–20 degrees *downwind* at the start. The feeling that a rocket ought to be directed into the wind, because it would otherwise be blown away, is completely wrong. If a rocket is pointed into the wind it will never reach its full height, but even that is something to be borne in mind because if the first rocket disappeared into low cloud, any subsequent rockets should be deliberately kept away from the vertical.

Hand-held flares have their idiosyncrasies too. Most will shed a little red-hot dross as they burn, so they must be held well away from the body – or the life raft – and downwind.

From time to time clubs organize demonstrations of firing pyrotechnics and they are well worth while. Although firing methods are at last becoming standardized there are still different makes in use and some – unfortunately – have a time lag built into the firing mechanism. There is no problem if the time lag is expected but there can be a very real one if the flare is thought to have misfired.

Remember when using a hand flare that the flame is intensely bright and, especially in the case of a white flare, all night vision will be completely ruined for a minute or more immediately after it has been fired.

The 'trick' here is to ensure that the method of firing is properly understood. Fire with the arm held well to leeward *but with the face turned away and the eyes very tightly shut*. You will be aware when the flare is extinguished and the handle can then be thrown over the side. By this method most, if not all, of that precious commodity – night vision – can be preserved.

Night vision

Pyrotechnics may cause night vision problems on occasions and for only a minute or so. Bad compass lighting or the careless use of cabin lights can cause the loss of proper vision for the look-out nearly all the time.

The subject is complex but the most important signal in the world may not be seen if the look-out cannot see properly.

Navigation lights must be mounted so that their rays cannot shine on light surfaces. Stanchions or pulpits can sometimes be treated with black tape to eliminate reflections. Compass and instrument lights *must* be fitted with dimmer switches – preferably a separate switch for each instrument. Recent research by the Institute of Aviation Medicine at Farnborough suggests that for instruments like a compass, where figures have to be read and the information passed to the brain, the light is best white, so long as it is dim enough. Apparently the brain takes longer to focus on a signal in coloured light, so that *very dim* white is better for both compass and chart light than the more usual red.

For cabin lighting, on the other hand, a very, very dim glow of deep red light throughout the cabin can be a joy. The watch below can see to dress or drink a cup of tea and, if the installation has been well planned, they have virtually full night vision immediately they reach the cockpit.

The nerves in the human eye may seem a far cry from the main subject of signalling, but if the man on watch cannot see properly the whole system of sea signalling is a risk.

There are two nervous systems within the eye. One is very sensitive to and easily affected by light, and this the brain needs to absorb information. The other, which the brain uses to distinguish movement and shadows, is largely unaffected by deep red light. If, therefore, all stray white light can be eliminated and a little red used below, the new watch can come on deck and see! With such organization you can 'see in the dark'. Fit separate dimmers to all instruments, and even ask any pipe smoker to warn those in the cockpit to close their eyes for a moment before he strikes his match. Use deck floodlights only for essential movements.

The environment on a small boat is difficult enough by day, in bad weather, but at night it can be almost impossible. Modern man is relying increasingly on instruments to tell him where he is and where he ought to go. To a certain extent instruments also tell him where other people are, and what they want to do. But when the environment is most unfriendly – when things go a bit wrong – it is basic signalling that we fall back on, and the basic means of communicating have not changed all that much for a couple of centuries.

Undoubtedly the magic of electronics will continue to allow us to do things our forefathers would never have dreamed of doing, but I have a feeling that the basics are likely to remain with us for some time to come.

Appendix A

Useful abbreviations

ALRS	Admiralty List of Radio Signals
AM	Amplitude modulation
ATC	Air Traffic Control
CG	Coastguard
Ch.	Channel(s)
CROSSMA	Centre Régional Opérationnel de Surveillance et de Sauvetage pour la Manche
CROSSA	Centre Régional Opérationnel de Surveillance et de Sauvetage pour l'Atlantique
CRS	Coast radio station (operated by British Telecom in UK)
D/F	Direction finding
DSB	Double sideband
DST	Daylight Saving Time. Note: it can be confusing to use the expression British Summer Time (BST) because most neighbouring countries now change to a summer time that may differ from that used in the UK.
EPIRB	Emergency Position Indicating Radio Beacon (also called a PLB, Personal Locator Beacon)
ETA	Estimated time of arrival
ETD	Estimated time of departure
FM	Frequency modulated
GRT	Gross Registered Tonnes
h	hour(s)
HF	High frequency
Hz	Hertz
H+ ...	Commencing at ... minutes past the hour
H24	Continuous
ICAO	International Civil Aviation Organization
Inop	Inoperative
IMO	International Maritime Organization (until 1982 Inter-governmental Maritime Consultative Organization)
ITU	International Telecommunication Union
kHz	Kilohertz
MF	Medium frequency
MHz	Megahertz
min	minute(s)
MRCC	Maritime Rescue Co-ordination Centre
MRSC	Maritime Rescue Sub-Centre
RNLI	Royal National Lifeboat Institution
R/T	Radiotelephony
Rx	Receiver
RYA	Royal Yachting Association
s	second(s)

SAR	Search and rescue
SITREP	Situation report
SSB	Single sideband
temp inop	Temporarily inoperative
Tx	Transmitter
ufn	Until further notice
UHF	Ultra high frequency
VHF	Very high frequency
VTM	Vessel Traffic Management
VTS	Vessel Traffic Services
W/T	Wireless telegraphy
Wx	Weather

Appendix B
VHF R/T anomalies
United States
In the United States Ch. 07, 18, 19, 21, 22, 23, 65, 66, 78, 79, 80, 83 and 88 operate in the simplex mode using the ship station frequencies, and 21, 22, 23 and 83 have the letters CG as a suffix; the remainder have the letter A as a suffix. American-made VHF and sets made elsewhere are thus not always compatible.

France and Belgium
Six special frequencies – Ch. 01, 03, 21, 23, 63 and 83 – were scheduled to have been phased out by January 1983.

The Netherlands
Some private small craft are not licensed by the Dutch authorities to use the interleaved Ch. 60–88. However, Ch. 77 is licensed in the Netherlands to be used – on low power only – as a social channel.

United Kingdom
In the UK Ch. 00 (zero) is used by the Royal National Lifeboat Institution, HM Coastguard, the Marine Police and the Services rescue aircraft as an SAR private channel. A special licence is required from the Home Office before Ch. 00 may be used and it is not normally granted to yachtsmen other than to a few Auxiliary Coastguards.

Also in the UK Ch. 67 is designated as a channel for direct communications between small craft and HM Coastguard 'for matters affecting the safety of the vessel'. The call is on Ch. 16. You will be asked to change to Ch. 67 for working.

Calling Coast Radio Stations (1983)
As is explained briefly in Chapter 5, the method of calling a coast radio station (CRS) on VHF radiotelephony varies from country to country. In the United Kingdom (except for Niton) all calls are made on Ch. 16. Niton Radio has changed to a system where they can accept calls on a working channel and it is planned that other UK CRS will follow.

If calling Niton, listen to each of the working channels, in turn, until one is found to be free. A channel in use will be indicated either by speech or by an engaged signal (a series of pips). A free channel carries no transmission.

France
The French CRS do not monitor Ch. 16; it is reserved for calls from coast stations. To call a French CRS choose the appropriate working channel, press the transmit

key for at least three seconds, and then listen. An exchange ringing tone should be heard until the CRS replies by asking for the name of the calling station.

Belgium
Calls are accepted either by calling on Ch. 16 or on the appropriate working channel. The call, from the coast, is always 'Oostende Radio'.

The Netherlands
Calls are accepted after calling on Ch. 16, but the Dutch prefer calls on the appropriate working channel. The call is always 'Scheveningen Radio' and should last at least three seconds. When no operator is immediately available, a chime is heard to indicate that the call has been accepted and will be attended to. Do not call again for about three minutes.

Federal Republic of Germany
Calls are accepted either on Ch. 16 or direct on the appropriate working channel.

Denmark, Norway, Sweden, Finland
Call on the appropriate working channels.

Poland
Call on Ch. 16.

German Democratic Republic
Changed in 1982 to accepting calls on working frequencies.

From that brief summary it is obvious that there are few hard and fast rules.
The details of the appropriate working frequencies for all the UK's immediate neighbours are given in the Admiralty *List of Radio Services for Small Craft* and in Macmillan's *Nautical Almanac*. Beyond Brest or the Elbe consult Admiralty *List of Radio Signals*, Vol. 1, Part 1.
If in any doubt listen, both on Ch. 16 and on the station's working frequency. The method preferred will usually become obvious, but do not confuse the situation by repeated calling. As a rule this simply 'pollutes' the air, as did the yacht that called an Irish CRS for six hours during the 1979 Fastnet Race, despite the fact that the Irish stations did not work on VHF frequencies at that time!

Appendix C
Citizens' Band R/T

Citizens' Band radio in some parts of the world, particularly the United States, has become a very big business, but the use of CB 'walkie-talkies' has been illegal in some European countries until recently.

In the United Kingdom CB radio was legalized towards the end of 1981 but the mode of emission was not compatible with most other foreign manufactured equipment and therefore the many illegal sets that had been imported into the United Kingdom are still illegal.

Licences for CB are obtainable from any Post Office in the UK (£10 in 1982) and the available channels are in two groups. Most of the sets now coming on the market are in the 27 MHz band, as used in the United States, but the important difference is that not only are our actual frequencies slightly different, but the British licence is for frequency modulated (FM) equipment, whereas the US sets use amplitude modulated (AM).

As is explained in Chapter 13, CB is potentially a most useful communications tool for yacht race management. However, as there is already evidence that at peak times – weekends – the 27 MHz channels are likely to suffer from 'cowboy' behaviour, it is suggested that the 934 MHz band, which will need slightly more expensive equipment and will have a strictly limited range, may well prove the answer for short-range yacht race management.

There are 40 channels in the UK in the 27 MHz band . . .

Channel number	Frequency	Channel number	Frequency
1	27.60125	21	27.80125
2	27.61125	22	27.81125
3	27.62125	23	27.82125
4	27.63125	24	27.83125
5	27.64125	25	27.84125
6	27.65125	26	27.85125
7	27.66125	27	27.86125
8	27.67125	28	27.87125
9	27.68125	29	27.88125
10	27.69125	30	27.89125
11	27.70125	31	27.90125
12	27.71125	32	27.91125
13	27.72125	33	27.92125
14	27.73125	34	27.93125
15	27.74125	35	27.94125
16	27.75125	36	27.95125
17	27.76125	37	27.96125
18	27.77125	38	27.97125
19	27.78125	39	27.98125
20	27.79125	40	27.99125

... and 20 channels in the 934 MHz band. Both use FM equipment.

Channel number	Frequency	Channel number	Frequency
1	934.025	11	934.525
2	934.075	12	934.575
3	934.125	13	934.625
4	934.175	14	934.675
5	934.225	15	934.725
6	934.275	16	934.775
7	934.325	17	934.825
8	934.375	18	934.875
9	934.425	19	934.925
10	934.475	20	934.975

Apart from the initial Post Office licence there are few restrictions on the use of CB sets: a lack of restriction is one of the reasons for having it.

In contrast to the practice with maritime VHF channels it is legal to use CB intership or ship-to-shore, or both.

In the 27 MHz band aerials must not be mounted more than 10 m above the ground or, if they are, the output must be reduced by 10 dB. In the 934 MHz band a set with an internal aerial is limited to 3 W effective radiated power. Otherwise the maximum r.f. output power and the effective radiated power on 27 MHz is 4 W and 2 W, and on 934 MHz it is 8 W and 25 W.

Appendix D
Medium frequency radiotelephony

The tremendous growth in the use of VHF R/T has almost swamped the small interest in MF R/T, but, as is mentioned in Part 1, the very strict type-approval standards applying to equipment to be fitted to vessels that are required by statute to carry MF radiotelephones were relaxed in 1981 for vessels voluntarily equipped. Thus small craft may install MF single-sideband R/T at an appreciably lower cost.

Nevertheless, even the cheapest MF equipment on the British market is still 10 to 15 times as expensive as VHF R/T; the sets still need a considerably more sophisticated installation, they must have a really good earth, and all but the very latest (1982) consume 20 A, or more, on transmit. Against these disadvantages is the overwhelming advantage that even the smallest MF R/T can be expected to have a range of 150 to 250 miles – sometimes more – and thus it is essential for mariners who want to be able to keep in touch with the shore when outside the 30 to 50 mile range of VHF R/T.

Combined medium frequency and high frequency sets are also available to a less sophisticated specification (although still type-approved by the Home Office). In round figures the new sets save 30 to 50 per cent of the cost of the larger and more complex equipment. They have ranges up to 1000 miles or more.

The principal difference in the specification is that the new 'small craft' sets do not have such stringent humidity and temperature control standards to meet. It could be argued therefore that if equipping a yacht for, say, a world cruise, the extra cost of the equipment built to the higher specification would not be all that great as a proportion of the overall cost, but the fact remains that there is now a far wider choice than there used to be.

Typical units to the new 'small craft' specification have a frequency range from 2 to 9 MHz or 2 to 18 MHz with 21 simplex (or one simplex and 19 duplex) channels in the smaller set, and 36 simplex (or 12 simplex and 12 duplex) in the larger.

Until recently MF R/T equipment was somewhat more complicated to operate when compared with the extreme simplicity of VHF R/T, but recently, synthesized channel selection and automatic switching of mode has changed all that; the latest MF equipment is almost as simple as VHF to use.

Anyone who might consider buying the older double-sideband MF equipment on some 'bargain offer' should note that the period of grace for the use of DSB has expired. It is now illegal – except on 2182 kHz and then only in emergency – to transmit on double sideband. In other words DSB sets are now obsolete.

Appendix E
Applications for the licences for radio telephone apparatus on small vessels

Ship Licence

After consulting a manufacturer of approved equipment, or his agent, about cost, supply and maintenance, apply for a licence from the Home Office, Radio Regulatory Department, Waterloo Bridge House, Waterloo Road, London SE1 8UA.

The Ship Licence includes the call sign, when the vessel is first licensed for radio, and remains with the vessel irrespective of change of ownership.

A Selective Call number will also be allocated if requested. The annual charge for a Ship Licence is £17.50 (1982).

The application form contains space for the details of the equipment to be used. It is a condition of the licence that radio apparatus installed shall comply with the various Performance Specifications issued. Full details of publications in the MPT series are available from the Home Office if necessary.

Certificate of Competency and the Authority to Operate

Certificates of Competency are awarded to anyone of any nationality provided the relevant examination has been passed. However, before operating a radio on United Kingdom registered ships the holder of the Certificate of Competency must also have an 'Authority to Operate' and they are generally restricted to British subjects and citizens of the Irish Republic. The Authority to Operate is normally issued with the Certificate of Competency.

If using VHF only the Certificate of Competency is marked 'Restricted (VHF Only)' and in these circumstances the equipment is not normally inspected on installation, as it is for MF sets, but any craft fitted with radio may be inspected at any time on a spot check basis.

To apply, write to British Maritime Radio Services, Landsec House, 23 New Fetter Lane, London EC4 1AE. The cost of the examination is £20 (1982), but see also Chapter 2.

Appendix F
Documents to be carried

Ship stations voluntarily fitted with radiotelephone equipment are required to carry the following:

The Ship Licence

A Copy of Section 11 of the Post Office (Protection) Act 1884

The Operator's Certificates of Competency in Radiotelephony and the *Authority to Operate*

A list of all coast radio stations with which communications are likely to be conducted

The Post Office Handbook for Radio Operators

A complete file of current Notices to Ship Wireless Stations:

Sent to all British Registered Ships or obtainable from British Telecom International, Maritime Radio Services Division, Landsec House, 23 New Fetter Lane, London EC4 1AE. Telephone 01-583 9416.

Appendix G
Signalling: Rules 4 and 5 from the International Yacht Racing Rules.
4 Signals
4.1 Visual signals
Unless otherwise prescribed by the national authority or in the sailing instructions, the following International Code flags and other visual signals shall be used as indicated and when displayed alone shall apply to all classes, and when displayed over a class signal they shall apply to the designated class only:

"AP"—Answering pendant—Postponement signal.
Means:
(a) *All races not started are* postponed. *The warning signal will be made one minute after this signal is lowered.*
   (One sound signal shall be made with the lowering of the AP.)
(b) Over one ball or shape.
   The scheduled starting times of all races not started are postponed *fifteen minutes.*
   (This *postponement* can be extended indefinitely by the addition of one ball or shape for every fifteen minutes.)
(c) Over one of the numeral pendants 1 to 9.
   All races not started are postponed *one hour, two hours, etc.*
(d) Over Code flag A.
   All races not started are postponed *to a later day.*

"B"—Protest signal.
When displayed by a yacht.
Means: I intend to lodge a protest.

"I"—Round the ends starting rule.
Broken out one minute before the starting signal is made, accompanied by one long sound signal. (This signal is used only when it is prescribed in the sailing instructions.)
Means: The one-minute period of the Round the Ends Starting Rule 51.1(c) has commenced.

"L"—*Means:*
(a) When displayed ashore:
   A notice to competitors has been posted on the notice board.
(b) When displayed afloat:
   Come within hail, or *Follow me.*

"M"—Mark signal.
When displayed on a buoy, vessel, or other object.
Means: Round or pass the object displaying this signal instead of the mark *which it replaces.*

"N"—Abandonment signal.
Means: All races are abandoned.

"N over X"—Abandonment and Re-sail Signal.
Means: All races are abandoned *and will shortly be resailed. The warning signal will be made one minute after this signal is lowered.*
(One sound signal shall be made with the lowering of "N over X".)

"N over First Substitute"—Cancellation signal.
Means: All races are cancelled.

"P"—Preparatory signal.
Means: The class designated by the warning signal will start *in five minutes exactly.*

"S"—Shorten course signal.
Means:
(a) at or near the starting line:
   Sail the shortened course prescribed in the sailing instructions.
(b) at or near the starting line:
   Finish *the race either:*
   (i) *at the prescribed finishing line at the end of the round still to be completed by the leading yacht,* or
   (ii) *in any other manner prescribed in the sailing instructions under Rule 3.2(a)(vi).*
(c) at or near a rounding *mark*:
   Finish *between the nearby* mark *and the committee boat.*

"X"—Individual recall.
Broken out immediately after the starting signal is made, accompanied by one sound signal, in accordance with Rule 8.2(b)(ii) (Recalls).
Means: One or more yachts have started prematurely or have infringed the Round the Ends Starting Rule 51.1(c).

"Y"—Life jacket signal.
Means: Life jackets or other adequate personal buoyancy shall be worn while racing by *all helmsmen and crews, unless specifically excepted in the sailing instructions.*
When this signal is displayed after the warning signal is made, failure to comply shall not be cause for disqualification.
Notwithstanding anything in this rule, it shall be the individual responsibility of each competitor to wear a life jacket or other adequate personal buoyancy when conditions warrant. A wet suit is not adequate personal buoyancy.

"First substitute"—General recall signal.
Means: The class is recalled for a new start as provided in the sailing instructions.
Unless the sailing instructions prescribe some other signal, the warning signal will be made one minute after this signal is lowered.
(One sound signal shall be made with the lower of First Substitute.)

Red flag—Displayed by committee boat.
Means: Leave all marks to port.

Green flag—Displayed by committee boat.
Means: Leave all marks to starboard.

Blue flag or shape—Finishing signal.
When displayed by a committee boat.
Means: The committee boat is on station at the finishing line.

4.2 Signalling the course
Unless otherwise prescribed by the national authority, the race committee shall either make the appropriate signal or otherwise designate the course before or with the warning signal.

4.3 Changing the course
The course for a class which has not started may be changed:
(a) when the only change is that a starting *mark* is to be shifted, by shifting the *mark* before the preparatory signal is made; or
(b) by displaying the appropriate *postponement* signal and indicating the new course before or with the warning signal to be displayed after the lowering of the *postponement* signal; or
(c) by displaying a course signal or by removing and substituting a course signal before or with the warning signal.

4.4 Signals for starting a race
(a) Unless otherwise prescribed by the national authority or in the sailing instructions, the signals for starting a race shall be made at five-minute intervals exactly, and shall be either

System 1 Warning signal   —Class flag broken out or distinctive signal displayed.
            Preparatory signal—Code flag "P" broken out or distinctive signal displayed.
            Starting signal   —Both warning and preparatory signals lowered.

In System 1 when classes are started:
 (i) at ten-minute intervals—
    the warning signal for each succeeding class shall be broken out or displayed at the starting signal of the preceding class,
(ii) at five-minute intervals—
    the preparatory signal for the first class to start shall be left displayed until the last class *starts*. The warning signal for each succeeding class shall be broken out or displayed at the preparatory signal of the preceding class,
or

System 2 Warning signal   —White or yellow shape.
            Preparatory signal—Blue shape.
            Starting signal for—Red shape.
              first class to start

In System 2 each signal shall be lowered one minute before the next is made. Class flags when used shall be broken out not later than the preparatory signal for each class.

In starting a series of classes:
  (i) at ten-minute intervals—
      the starting signal for each class shall be the warning signal for the next.
  (ii) at five-minute intervals—
      the preparatory signal for each class shall be the warning signal for the next.
(b) Although rules 4.1 "**P**" and 4.4(a) specify five-minute intervals between signals, this shall not interfere with the power of a race committee to start a series of races at any intervals which it considers desirable.
(c) A warning signal shall not be made before its scheduled time, except with the consent of all yachts entitled to *race*.
(d) When a significant error is made in the timing of the interval between any of the signals for starting a race, the recommended procedure is to signal a general recall, *postponement* or *abandonment* of the race whose start is directly affected by the error and a corresponding *postponement* of succeeding races. Unless otherwise prescribed in the sailing instructions a new warning signal shall be made. When the race is not recalled, *postponed* or *abandoned* after an error in the timing of the interval, each succeeding signal shall be made at the correct interval from the preceding signal.

4.5 Other signals

The sailing instructions shall designate any other special signals and shall explain their meaning.

4.6 Calling attention to signals

Whenever the race committee makes a signal, except "**S**" before the warning signal or a blue flag or shape when on station at the finishing line, it shall call attention to its action as follows:
(a) Three guns or other sound signals when displaying:
     (i) "**N**";
     (ii) "**N over X**";
     (iii) "**N over First Substitute**".
(b) Two guns or other sound signals when displaying:
     (i) "**AP**";
     (ii) "**S**";
     (iii) "**First Substitute**".
(c) One gun or other sound signal when making any other signal, including the lowering of:
     (i) "**AP**" when the length of postponement is not signalled;
     (ii) "**N over X**";
     (iii) "**First Substitute**".

4.7 Visual starting signals to govern

Times shall be taken from the visual starting signals, and a failure or mistiming of a gun or other sound signal calling attention to starting signals shall be disregarded.

5 Postponing, abandoning or cancelling a race and changing or shortening course

5.1

The race committee

(a) before the starting signal may shorten the course or *postpone* or *cancel* a race for any reason.
(b) after the starting signal may shorten the course by finishing a race at any rounding *mark* or *abandon* or *cancel* a race because of foul weather endangering the yachts, or because of insufficient wind, or because a *mark* is missing or has shifted for other reasons directly affecting safety or the fairness of the competition.
(c) after the starting signal may change the course at any rounding *mark* subject to proper notice being given to each yacht as prescribed in the sailing instructions.
(d) after a race has been completed, shall not *abandon* or *cancel* it without taking the appropriate action under rule 74.2 (Consideration of Redress).

5.2

After a *postponement* the ordinary starting signals prescribed in rule 4.4(a) (Signals), shall be used, and the postponement signal, when a general one, shall be lowered one minute before the first warning or course signal is made.

5.3

The race committee shall notify all yachts concerned by signal or otherwise when and where a race *postponed* or *abandoned* will be sailed.

Appendix H
United States inland rules: Signalling differences

In the United States, new Inland Rules have recently been introduced and, broadly, they are much more in line with the International Regulations for Preventing Collision at Sea than were the earlier regulations.

One important point is that vessels of less than 20 m (65 ft) l.o.a. that complied with the previous rules are not required to change their lights to comply with the new, increased ranges.

Another point of interest is that the category of small power-driven vessels of less than 7 m (23 ft) and capable of less than 7 knots, does not appear in the US Inland Rules.

The somewhat controversial Rule 25 in the Collision Regulations, which requires vessels under sail to exhibit a black cone when they are also being propelled by machinery, is not applicable in the new US Inland Rules to craft under 12 m (40 ft) l.o.a.

Finally there is an additional signal for vessels pushing a tow. It is designated a 'special flashing light'. It flashes yellow, at from 50 to 70 times per minute, and is exhibited from the bow of the vessel being pushed ahead.

Generally the new rules have been welcomed by US yachtsmen because, if nothing else, they have eliminated the ambiguity of the old Rule 17 affecting sailing craft giving way to each other, which was still based on the pre-1960 Collision Regulations.

Appendix I

Conversion factors

Length
1 m = 39.3701 in
1 m = 3.2808 ft
1 m = 0.547 fathom
1 km = 0.540 n mile

Length
1 in = 0.0254 m
1 ft = 0.3048 m
1 fathom = 1.83 m
1 n mile = 1.85 km

Area
1 mm² = 0.00155 in²
1 m² = 10.764 ft²
1 hectare = 2.47 acres

Area
1 in² = 645 mm²
1 ft² = 0.093 m²
1 acre = 0.405 hectare

Volume
1 m³ = 35.315 ft³
1 litre = 0.22 gall

Volume
1 ft³ = 0.028 m³
1 gall = 4.546 litre

Speed
1 km/h = 0.54 knot
1 km/h = 0.62 mile/h

Speed
1 mile/h = 1.61 km/h
1 knot = 1.85 km/h
1 knot = 1.15 mile/h

Mass
1 kg = 2.205 lb
1 tonne = 0.984 ton

Mass
1 lb = 0.454 kg
1 ton = 1.016 tonne

Pressure
1000 millibars = 29.53 in (Hg)

Pressure
1 in (Hg) = 33.86 millibars*

Temperature
°C = 0.556 (F−32)

Temperature
F = 1.8 °C+32

Force
1 Newton (N) = 0.225 lbf
1 kN = 0.100 tonf

Force
1 lbf = 4.45 (N)
1 tonf = 9.96 kN

Torque
(Moment of force)
1 Nm = 0.7376 lbf/ft
1 kgf/m = 0.1383 lb/ft

Torque
(Moment of force)
1 lbf/ft = 1.3558 Nm
1 lb/ft = 7.2330 kgf/m

Energy
1 Joule (J) = 0.738 ft lbf
1 kJ = 0.948 Btu

Energy
1 ft lbf = 1.36 Joule (J)
1 Btu = 1.06 kJ

Power
1 kW = 1.34 hp

Power
1 hp = 0.746 kW

*See also Appendix J.

Appendix J
Metric units and their symbols and other useful measurements

Quantity	Name	Symbol and/or explanation
length	metre	m
	nautical mile	n mile or M, as in abbreviation for a light: Iso 4 s 43 m 18 M.
	cable	One tenth of a nautical mile.
mass	kilogram	kg
	tonne	Written 'tonne' to avoid confusion with short or long tons of American or Imperial measure. Do not abbreviate as 't' to avoid confusion with 'l' or '1'.
time	year	Abbreviated to last two digits: '82.
	month	Abbreviated to first three letters.
	day(s)	d
	hour(s)	h
	minute(s)	min In a tabulation minutes can be shown as 3 h 45 as in tidal information, or as H+20. Otherwise time is given in four-figure notation: 1435. Assume GMT unless marked otherwise.
	second(s)	s
plane angle	degree	Bearings in three-figure notation. True unless stated otherwise.
electric current	ampere	A
power	watt	W
electrical potential	volt	V The derivation is W/A.
electrical resistance	ohm	Ω The derivation is V/A.
frequency	Hertz	Hz The derivation is one cycle per second.
temperature	degree Celsius	°C
luminous/intensity	candela	cd
area	square metre	m^2
	hectare	ha Equals 10 000 m^2

Quantity	*Name*	*Symbol and/or explanation*
volume	cubic metre	m³
	litre	Always write litre, not l, because a small letter 'l' will be confused with the figure 1.
speed	knot	kn Nautical mile per hour.
	metres per second	m/s The quantity normally used by meteorologists.
	kilometres per hour	km/h The quantity normally used in aviation.
pressure	millibar	mb By 1985 the bar is likely to be replaced by the Pascal in weather forecasts. However as one hectopascal equals one millibar the figures will be the same even if it is kilopascals that are used as the unit; hecto is not a preferred SI prefix.

Prefixes

mega	one million times	M
kilo	one thousand times	k
milli	one thousandth	m
micro	one millionth	μ (The Greek letter, not a small u)

197

Appendix K
How to group digits and to use numbers with units
Where a number is less than one, write 0.25 m; not .25 m.

Measurement should only use one unit. Write 1.24 m or 124 cm; not 1 m 24 cm.

When grouping digits four should be written without a space: 4242, but five or more should be grouped in blocks of three: 3 000 000. This rule should be applied both sides of a decimal marker (a comma in some countries): 4 321.123 4.

However, in tabulation digits may be grouped in threes;
2 000
23 445
123 554

When speaking numbers, such as for a telephone code and number, the station called will not know in advance how many digits there will be. It is far easier to write down a number correctly if it is spoken in groups of two digits at a time, *paired off from the right*: 'One – two three – four five' not 'One two – three four – five'.

The reason is that by starting with a single digit, when the total is an odd number, the station called then knows that all that follows will be in pairs of digits.

Note: The *Standard Marine Navigation Vocabulary* recommends that all digits (except for the round thousand) should be spoken; 'Four zero zero one' is far more likely to be taken down correctly than 'Four thousand and one'.

Appendix L
Signal lights
The International Standards Organization has recommended the following for signal lights on board ships:

red	alarm, dangerous conditions
amber	warning of anomalous conditions
green	normal operating condition
blue	instructions, information
white	general information

As a general indication flashing lights should be reserved for situations requiring immediate action.

Appendix M
International regulations for preventing collisions at sea, 1972

Complete with the amendments adopted by the International Maritime Organization which are in force from June 1983 are indicated by a side line: thus

Rule 3: General definitions

For the purpose of these Rules, except where the context otherwise requires:

(a) The word 'vessel' includes every description of water craft, including non-displacement craft and seaplanes, used or capable of being used as a means of transportation on water.

(b) The term 'power-driven vessel' means any vessel propelled by machinery.

(c) The term 'sailing vessel' means any vessel under sail provided that propelling machinery, if fitted, is not being used.

(d) The term 'vessel engaged in fishing' means any vessel fishing with nets, lines, trawls or other fishing apparatus which restrict manoeuvrability, but does not include a vessel fishing with trolling lines or other fishing apparatus which do not restrict manoeuvrability.

(e) The word 'seaplane' includes any aircraft designed to manoeuvre on the water.

(f) The term 'vessel not under command' means a vessel which through some exceptional circumstance is unable to manoeuvre as required by these Rules and is therefore unable to keep out of the way of another vessel.

(g) The term 'vessel restricted in her ability to manoeuvre' means a vessel which from the nature of her work is restricted in her ability to manoeuvre as required by these Rules and is therefore unable to keep out of the way of another vessel.

The term 'vessels restricted in their ability to manoeuvre' shall include but not be limited to:

- (i) a vessel engaged in laying, servicing or picking up a navigation mark, submarine cable or pipeline;
- (ii) a vessel engaged in dredging, surveying or underwater operations;
- (iii) a vessel engaged in replenishment or transferring persons, provisions or cargo while under way;
- (iv) a vessel engaged in the launching or recovery of aircraft;
- (v) a vessel engaged in mineclearance operations;
- (vi) a vessel engaged in a towing operation such as severely restricts the towing vessel and her tow in their ability to deviate from their course.

(h) The term 'vessel constrained by her draught' means a power-driven vessel which because of her draught in relation to the available depth of water is severely restricted in her ability to deviate from the course she is following.

(i) The word 'underway' means that a vessel is not at anchor, or made fast to the shore, or aground.

(j) The words 'length' and 'breadth' of a vessel mean her length overall and greatest breadth.

(k) Vessels shall be deemed to be in the sight of one another only when one can be observed from the other.

(l) The term 'restricted visibility' means any condition in which visibility is restricted by fog, mist, falling snow, heavy rainstorms, sandstorms or any other similar causes.

Part C: Lights and shapes
Rule 20: Application
(a) Rules in this Part shall be complied with in all weathers.
(b) The Rules concerning lights shall be complied with from sunset to sunrise, and during such times no other lights shall be exhibited, except such lights as cannot be mistaken for the lights specified in these Rules or do not impair their visibility or distinctive character, or interfere with the keeping of a proper look-out.
(c) The lights prescribed by these Rules shall, if carried, also be exhibited from sunset to sunrise in restricted visibility and may be exhibited in all other circumstances when it is deemed necessary.
(d) The Rules concerning shapes shall be complied with by day.
(e) The lights and shapes specified in these Rules shall comply with the provisions of Annex I to these Regulations.

Rule 21: Definitions
(a) 'Masthead light' means a white light placed over the fore and aft centreline of the vessel showing an unbroken light over an arc of horizon of 225 degrees and so fixed as to show the light from right ahead to 22.5 degrees abaft the beam on either side of the vessel.
(b) 'Sidelights' means a green light on the starboard side and a red light on the port side each showing an unbroken light over an arc of the horizon of 112.5 degrees and so fixed as to show the light from right ahead to 22.5 degrees abaft the beam on its respective side. In a vessel of less than 20 m length the sidelights may be combined in one lantern carried on the fore and aft centreline of the vessel.
(c) 'Sternlight' means a white light placed as nearly as practicable at the stern showing an unbroken light over an arc of the horizon of 135 degrees and so fixed as to show the light 67.5 degrees from right aft on each side of the vessel.
(d) 'Towing light' means a yellow light having the same characteristics as the 'sternlight' defined in paragraph (c) of this Rule.
(e) 'All-round light' means a light showing an unbroken light over an arc of the horizon of 360 degrees.
(f) 'Flashing light' means a light flashing at regular intervals at a frequency of 120 flashes or more per minute.

Rule 22: Visibility of lights
The lights prescribed in these Rules shall have an intensity as specified in Section 8 of Annex I to these Regulations so as to be visible at the following minimum ranges:
(a) In vessels of 50 metres or more in length:
 —a masthead light, 6 miles;

—a sidelight, 3 miles;
—a sternlight, 3 miles;
—a towing light, 3 miles;
—a white, red, green or yellow all-round light, 3 miles.
- **(b)** In vessels of 12 metres or more in length but less than 50 metres in length:
—a masthead light, 5 miles; except that where the length of the vessel is less than 20 metres, 3 miles;
—a sidelight, 2 miles;
—a sternlight, 2 miles;
—a towing light, 2 miles;
—a white, red, green or yellow all-round light, 2 miles.
- **(c)** In vessels of less than 12 metres in length:
—a masthead light, 2 miles;
—a sidelight, 1 mile;
—a sternlight, 2 miles;
—a towing light, 2 miles;
—a white, red, green or yellow all-round light, 2 miles.
- **(d)** In inconspicuous, partly submerged vessels or objects being towed:
—a white all-round light, 3 miles.

Rule 23: Power-driven vessels underway

- **(a)** A power-driven vessel underway shall exhibit:
    - (i) a masthead light forward;
    - (ii) a second masthead light abaft of and higher than the forward one; except that a vessel of less than 50 metres in length shall not be obliged to exhibit such light but may do so;
    - (iii) sidelights;
    - (iv) a sternlight.
- **(b)** An air-cushion vessel when operating in the non-displacement mode shall, in addition to the lights prescribed in paragraph(*a*) of this Rule, exhibit an all-round flashing yellow light.
- **(c)** (i) A power-driven vessel of less than 12 metres in length may in lieu of the lights prescribed in paragraph (*a*) of this Rule exhibit an all-round white light and sidelights;
    - (ii) a power-driven vessel of less than 7 metres in length whose maximum speed does not exceed 7 knots may in lieu of the lights prescribed in paragraph (*a*) of this Rule exhibit an all-round white light and shall, if practicable, also exhibit sidelights;
    - (iii) the masthead light or all-round white light on a power-driven vessel of less than 12 metres in length may be displaced from the fore and aft centreline of the vessel if centreline fitting is not practicable, provided that the sidelights are combined in one lantern which shall be carried on the fore and aft centreline of the vessel or located as nearly as practicable in the same fore and aft line as the masthead light or the all-round white light.

Rule 24: Towing and pushing

(a) A power-driven vessel when towing shall exhibit:
  (i) instead of the light prescribed in Rule 23(a)(i) or (a)(ii), two masthead lights in a vertical line. When the length of the tow, measuring from the stern of the towing vessel to the after end of the tow exceeds 200 metres, three such lights in a vertical line;
  (ii) sidelights;
  (iii) a sternlight;
  (iv) a towing light in a vertical line above the sternlight;
  (v) when the length of the two exceeds 200 metres, a diamond shape where it can best be seen.

(b) When a pushing vessel and a vessel being pushed ahead are rigidly connected in a composite unit they shall be regarded as a power-driven vessel and exhibit the lights prescribed in Rule 23.

(c) A power-driven vessel when pushing ahead or towing alongside, except in the case of a composite unit, shall exhibit:
  (i) instead of the light prescribed in Rule 23(a)(i) or (a)(ii), two masthead lights forward in a vertical line;
  (ii) sidelights;
  (iii) a sternlight.

(d) A power-driven vessel to which paragraph (a) or (c) of this Rule apply shall also comply with Rule 23(a)(ii).

(e) A vessel or object being towed, other than those mentioned in paragraph (g) of this Rule, shall exhibit:
  (i) sidelights;
  (ii) a sternlight;
  (iii) when the length of the tow exceeds 200 metres, a diamond shape where it can best be seen.

(f) Provided that any number of vessels being towed alongside or pushed in a group shall be lighted as one vessel:
  (i) a vessel being pushed ahead, not being part of a composite unit, shall exhibit at the forward end, sidelights;
  (ii) a vessel being towed alongside shall exhibit a sternlight and at the forward end, sidelights.

(g) An inconspicuous, partly submerged vessel or object, or combination of such vessels or objects being towed, shall exhibit:
  (i) if it is less than 25 metres in breadth, one all-round white light at or near the forward end and one at or near the after end except that dracones need not exhibit a light at or near the forward end;
  (ii) if it is 25 metres or more in breadth, two additional all-round white lights at or near the extremities of its breadth;
  (iii) if it exceeds 100 metres in length, additional all-round white lights between the lights prescribed in sub-paragraphs (i) and (ii) so that the distance between the lights shall not exceed 100 metres;
  (iv) a diamond shape at or near the aftermost extremity of the last vessel or object being towed and if the length of the tow exceeds 200 metres an additional diamond shape where it can best be seen and located as far forward as is practicable.

(h) Where from any sufficient cause it is impracticable for a vessel or object being towed to exhibit the lights or shapes prescribed in paragraph (*e*) or (*g*) of this Rule, all possible measures shall be taken to light the vessel or object being towed or at least to indicate the presence of such vessel or object.

(i) Where from any sufficient cause it is impracticable for a vessel not normally engaged in towing operations to display the lights prescribed in paragraph (*a*) or (*c*) of this Rule, such vessel shall not be required to exhibit those lights when engaged in towing another vessel in distress or otherwise in need of assistance. All possible measures shall be taken to indicate the nature of the relationship between the towing vessel and the vessel being towed as authorized by Rule 36, in particular to illuminate the towline.

Rule 25: *Sailing vessels underway and vessels under oars*

(a) A sailing vessel underway shall exhibit:
 (i) sidelights;
 (ii) a sternlight.

(b) In a sailing vessel of less than 20 metres in length the lights prescribed in paragraph (*a*) of this Rule may be combined in one lantern carried at or near the top of the mast where it can best be seen.

(c) A sailing vessel underway may, in addition to the lights prescribed in paragraph (*a*) of this Rule, exhibit at or near the top of the mast, where they can best be seen, two all-round lights in a vertical line, the upper being red and the lower green, but these lights shall not be exhibited in conjunction with the combined lantern permitted by paragraph (*b*) of this Rule.

(d) (i) A sailing vessel of less than 7 metres in length shall, if practicable, exhibit the lights prescribed in paragraphs (*a*) or (*b*) of this Rule, but if she does not, she shall have ready at hand an electric torch or lighted lantern showing a white light which shall be exhibited in sufficient time to prevent collision.

 (ii) A vessel under oars may exhibit the lights prescribed in this Rule for sailing vessels, but if she does not, she shall have ready at hand an electric torch or lighted lantern showing a white light which shall be exhibited in sufficient time to prevent collision.

(e) A vessel proceeding under sail when also being propelled by machinery shall exhibit forward where it can best be seen a conical shape, apex downwards.

Rule 26: *Fishing vessels*

(a) A vessel engaged in fishing, whether underway or at anchor, shall exhibit only the lights and shapes prescribed in this Rule.

(b) A vessel when engaged in trawling, by which is meant the dragging through the water of a dredge net or other apparatus used as a fishing appliance, shall exhibit:
 (i) two all-round lights in a vertical line, the upper being green and the lower white, or a shape consisting of two cones with their apexes together in a vertical line one above the other; a vessel of less than 20 metres in length may instead of this shape exhibit a basket;

(ii) a masthead light abaft of and higher than the all-round green light; a vessel of less than 50 metres in length shall not be obliged to exhibit such a light but may do so;
  (iii) when making way through the water, in addition to the lights prescribed in this paragraph, sidelights and a sternlight.
(c) A vessel engaged in fishing, other than trawling, shall exhibit:
  (i) two all-round lights in a vertical line, the upper being red and the lower white, or a shape consisting of two cones with apexes together in a vertical line one above the other; a vessel of less than 20 metres in length may instead of this shape exhibit a basket;
  (ii) when there is outlying gear extending more than 150 metres horizontally from the vessel, an all-round white light or a cone apex upwards in the direction of the gear;
  (iii) when making way through the water, in addition to the lights prescribed in this paragraph, sidelights and a sternlight.
(d) A vessel engaged in fishing in close proximity to other vessels engaged in fishing may exhibit the additional signals described in Annex II to these Regulations.
(e) A vessel when not engaged in fishing shall not exhibit the lights or shapes prescribed in this Rule, but only those prescribed for a vessel of her length.

Rule 27: Vessels not under command or restricted in their ability to manoeuvre

(a) A vessel not under command shall exhibit:
  (i) two all-round red lights in a vertical line where they can best be seen;
  (ii) two balls or similar shapes in a vertical line where they can best be seen;
  (iii) when making way through the water, in addition to the lights prescribed in this paragraph, sidelights and a sternlight.
(b) A vessel restricted in her ability to manoeuvre, except a vessel engaged in mine clearance operations, shall exhibit:
  (i) three all-round lights in a vertical line where they can best be seen. The highest and lowest of these lights shall be red and the middle light shall be white;
  (ii) three shapes in a vertical line where they can best be seen. The highest and lowest of these shapes shall be balls and the middle one a diamond;
  (iii) when making way through the water, masthead light or lights, sidelights and a sternlight, in addition to the lights prescribed in sub-paragraph (i);
  (iv) when at anchor, in addition to the lights or shapes prescribed in sub-paragraphs (i) and (ii), the light, lights or shape prescribed in Rule 30.
(c) A power-driven vessel engaged in a towing operation such as severely restricts the towing vessel and her tow in their ability to deviate from their course shall, in addition to the lights or shapes prescribed in Rule 24(*a*), exhibit the lights or shapes prescribed in sub-paragraphs (*b*)(i) and (ii) of this Rule.
(d) A vessel engaged in dredging or underwater operations, when restricted in her ability to manoeuvre, shall exhibit the lights and shapes prescribed in sub-

paragraphs (*b*)(i), (ii) and (iii) of this Rule and shall in addition, when an obstruction exists, exhibit:
- (i) two all-round red lights or two balls in a vertical line to indicate the side on which the obstruction exists;
- (ii) two all-round green lights or two diamonds in a vertical line to indicate the side on which another vessel may pass;
- (iii) when at anchor, the lights or shapes prescribed in this paragraph instead of the lights or shape prescribed in Rule 30.

(e) Whenever the size of a vessel engaged in diving operations makes it impracticable to exhibit all lights and shapes prescribed in paragraph (*d*) of this Rule, the following shall be exhibited:
- (i) three all-round lights in a vertical line where they can best be seen. The highest and lowest of these lights shall be red and the middle light shall be white;
- (ii) a rigid replica of the International Code flag "A" not less than 1 metre in height. Measures shall be taken to ensure its all-round visibility.

(f) A vessel engaged in mineclearance operations shall in addition to the lights prescribed for a power-driven vessel in Rule 23 or to the lights or shape prescribed for a vessel at anchor in Rule 30 as appropriate, exhibit three all-round green lights or three balls. One of these lights or shapes shall be exhibited near the foremast head and one at each end of the fore yard. These lights or shapes indicate that it is dangerous for another vessel to approach within 1000 metres of the mine clearance vessel.

(g) Vessels of less than 12 metres in length, except those engaged in diving operations, shall not be required to exhibit the lights and shapes prescribed in this Rule.

(h) The signals prescribed in this Rule are not signals of vessels in distress and requiring assistance. Such signals are contained in Annex IV to these Regulations.

Rule 28: Vessels constrained by their draught

A vessel constrained by her draught may, in addition to the lights prescribed for power-driven vessels in Rule 23, exhibit where they can best be seen three all-round red lights in a vertical line, or a cylinder.

Rule 29: Pilot vessels

(a) A vessel engaged on pilotage duty shall exhibit:
- (i) at or near the masthead, two all-round lights in a vertical line, the upper being white and the lower red;
- (ii) when under way, in addition, sidelights and a sternlight;
- (iii) when at anchor, in addition to the lights prescribed in sub-paragraph (i), the light, lights or shape prescribed in Rule 30 for vessels at anchor.

(b) A pilot vessel when not engaged on pilotage duty shall exhibit the lights or shapes prescribed for a similar vessel of her length.

Rule 30: Anchored vessels and vessels aground
(a) A vessel at anchor shall exhibit where it can best be seen:
  (i) in the fore part, an all-round white light or one ball;
  (ii) at or near the stern and at a lower level than the light prescribed in sub-paragraph (i), an all-round white light.

(b) A vessel of less than 50 metres in length may exhibit an all-round white light where it can best be seen instead of the lights prescribed in paragraph (a) of this Rule.

(c) A vessel at anchor may, and a vessel of 100 metres and more in length shall, also use the available working or equivalent lights to illuminate her decks.

(d) A vessel aground shall exhibit the lights prescribed in paragraph (a) or (b) of this Rule and in addition, where they can best be seen:
  (i) two all-round red lights in a vertical line;
  (ii) three balls in a vertical line.

(e) A vessel of less than 7 metres in length, when at anchor, not in or near a narrow channel, fairway or anchorage, or where other vessels normally navigate, shall not be required to exhibit the lights or shapes prescribed in paragraphs (a) and (b) of this Rule.

(f) A vessel of less than 12 metres in length, when aground, shall not be required to exhibit the lights or shapes prescribed in sub-paragraph (d)(i) and (ii) of this Rule.

Rule 31: Seaplanes
Where it is impracticable for a seaplane to exhibit lights and shapes of the characteristics or in the position prescribed in the Rules of this Part she shall exhibit lights and shapes as closely similar in characteristics and position as is possible.

Part D: Sound and light signals

Rule 32: Definitions
(a) The word 'whistle' means any sound signalling appliance capable of producing the prescribed blasts and which complies with the specifications in Annex III to these Regulations.

(b) The term 'short blast' means a blast of about one second's duration.

(c) The term 'prolonged blast' means a blast of from four to six seconds' duration.

Rule 33: Equipment for sound signals
(a) A vessel of 12 metres or more in length shall be provided with a whistle and a bell and a vessel of 100 metres or more in length shall, in addition, be provided with a gong, the tone and sound of which cannot be confused with that of the bell. The whistle, bell and gong shall comply with the specifications in Annex III to these Regulations. The bell or gong or both may be replaced by other equipment having the same respective sound characteristics, provided that manual sounding of the prescribed signals shall always be possible.

(b) A vessel of less than 12 metres in length shall not be obliged to carry the sound signalling appliances prescribed in paragraph (a) of this Rule but if she does not, she shall be provided with some other means of making an efficient sound signal.

Rule 34: Manoeuvring and warning signals

(a) When vessels are in sight of one another, a power-driven vessel underway, when manoeuvring as authorized or required by these Rules, shall indicate that manoeuvre by the following signals on her whistle:
—one short blast to mean *I am altering my course to starboard*;
—two short blasts to mean *I am altering my course to port*;
—three short blasts to mean *I am operating astern propulsion*.

(b) Any vessel may supplement the whistle signals prescribed in paragraph (a) of this Rule by light signals, repeated as appropriate, whilst the manoeuvre is being carried out:
  (i) these light signals shall have the following significance:
    —one flash to mean *I am altering my course to starboard*;
    —two flashes to mean *I am altering my course to port*;
    —three flashes to mean *I am operating astern propulsion*;
  (ii) the duration of each flash shall be about one second, the interval between flashes shall be about one second, and the interval between successive signals shall not be less than ten seconds;
  (iii) the light used for this signal shall, if fitted, be an all-round white light, visible at a minimum range of 5 miles, and shall comply with the provisions of Annex I to these Regulations.

(c) When in sight of one another in a narrow channel or fairway:
  (i) a vessel intending to overtake another shall in compliance with Rule 9(e)(i) indicate her intention by the following signals on her whistle:
    —two prolonged blasts followed by one short blast to mean *I intend to overtake you on your starboard side*;
    —two prolonged blasts followed by two short blasts to mean *I intend to overtake you on your port side*;
  (ii) the vessel about to be overtaken when acting in accordance with Rule 9(e)(i) shall indicate her agreement by the following signal on her whistle:
    —one prolonged, one short, one prolonged and one short blast, in that order.

(d) When vessels in sight of one another are approaching each other and from any cause either vessel fails to understand the intentions or actions of the other, or is in doubt whether sufficient action is being taken by the other to avoid collision, the vessel in doubt shall immediately indicate such doubt by giving at least five short and rapid blasts on the whistle. Such signal may be supplemented by a light signal of at least five short and rapid flashes.

(e) A vessel nearing a bend or an area of channel or fairway where other vessels may be obscured by an intervening obstruction shall sound one prolonged blast. Such signal shall be answered with a prolonged blast by any approaching vessel that may be within hearing around the bend or behind the intervening obstruction.

(f) If whistles are fitted on a vessel at a distance apart of more than 100 metres, one whistle only shall be used for giving manoeuvring and warning signals.

Rule 35: Sound signals in restricted visibility

In or near an area of restricted visibility, whether by day or night, the signals prescribed in this Rule shall be used as follows:

(a) A power-driven vessel making way through the water shall sound at intervals of not more than 2 minutes one prolonged blast.

(b) A power-driven vessel underway but stopped and making no way through the water shall sound at intervals of not more than 2 minutes two prolonged blasts in succession with an interval of about 2 seconds between them.

(c) A vessel not under command, a vessel restricted in her ability to manoeuvre, a vessel constrained by her draught, a sailing vessel, a vessel engaged in fishing and a vessel engaged in towing or pushing another vessel shall, instead of the signals prescribed in paragraphs (*a*) or (*b*) of this Rule, sound at intervals of not more than 2 minutes three blasts in succession, namely one prolonged followed by two short blasts.

(d) A vessel engaged in fishing, when at anchor, and a vessel restricted in her ability to manoeuvre when carrying out her work at anchor, shall instead of the signals prescribed in paragraph (*g*) of this Rule sound the signal prescribed in paragraph (*c*) of this Rule.

(e) A vessel towed or if more than one vessel is towed the last vessel of the tow, if manned, shall at intervals of not more than 2 minutes sound four blasts in succession, namely one prolonged followed by three short blasts. When practicable, this signal shall be made immediately after the signal made by the towing vessel.

(f) When a pushing vessel and a vessel being pushed ahead are rigidly connected in a composite unit they shall be regarded as a power-driven vessel and shall give the signals prescribed in paragraphs (*a*) or (*b*) of this Rule.

(g) A vessel at anchor shall at intervals of not more than one minute ring the bell rapidly for about 5 seconds. In a vessel of 100 metres or more in length the bell shall be sounded in the fore part of the vessel and immediately after the ringing of the bell the gong shall be sounded rapidly for about 5 seconds in the after part of the vessel. A vessel at anchor may in addition sound three blasts in succession, namely one short, one prolonged and one short blast, to give warning of her position and of the possibility of collision to an approaching vessel.

(h) A vessel aground shall give the bell signal and if required the gong signal prescribed in paragraph (*f*) of this Rule and shall, in addition, give three separate and distinct strokes on the bell immediately before and after the rapid ringing of the bell. A vessel aground may in addition sound an appropriate whistle signal.

(i) A vessel of less than 12 metres in length shall not be obliged to give the above-mentioned signals but, if she does not, shall make some other efficient sound signal at intervals of not more than 2 minutes.

(j) A pilot vessel when engaged on pilotage duty may in addition to the signals prescribed in paragraphs (*a*), (*b*) or (*f*) of this Rule sound an identity signal consisting of four short blasts.

Rule 36: Signals to attract attention

If necessary to attract the attention of another vessel any vessel may make light or sound signals that cannot be mistaken for any signal authorized elsewhere in these

Rules, or may direct the beam of her searchlight in the direction of the danger, in such a way as not to embarrass any vessel.

Any light to attract the attention of another vessel shall be such that it cannot be mistaken for any aid to navigation. For the purpose of this Rule, the use of high intensity intermittent or revolving lights, such as strobe lights, shall be avoided.

Rule 37: *Distress signals*

When a vessel is in distress and requires assistance she shall use or exhibit the signals described in Annex IV to these Regulations.

Part E: Exemptions

Rule 38: *Exemptions*

Any vessel (or class of vessels) provided that she complies with the requirements of the International Regulations for Preventing Collisions at Sea, 1960, the keel of which is laid or which is at a corresponding stage of construction before the entry into force of these Regulations may be exempted from compliance therewith as follows:

(a) The installation of lights with ranges prescribed in Rule 22, until four years after the date of entry into force of these Regulations.

(b) The installation of lights with colour specifications as prescribed in Section 7 of Annex I to these Regulations, until four years after the date of entry into force of these Regulations.

(c) The repositioning of lights as a result of conversion from Imperial to metric units and rounding off measurement figures, permanent exemption.

(d) (i) The repositioning of masthead lights on vessels of less than 150 metres in length, resulting from the prescriptions of Section 3(*a*) of Annex I to these Regulations, permanent exemption.

(ii) The repositioning of masthead lights on vessels of 150 metres or more in length, resulting from the prescriptions of Section 3(*a*) of Annex I to these Regulations, until nine years after the date of entry into force of these Regulations.

(e) The repositioning of masthead lights resulting from the prescriptions of Section 2(*b*) of Annex I of these Regulations, until nine years after the date of entry into force of these Regulations.

(f) The repositioning of sidelights resulting from the prescriptions of Sections 2(*g*) and 3(*b*) of Annex I to these Regulations, until nine years after the date of entry into force of these Regulations.

(g) The requirements for sound signal appliances prescribed in Annex III to these Regulations, until nine years after the date of entry into force of these Regulations.

(h) The repositioning of all-round lights resulting from the prescription of Section 9(*b*) of Annex I to these Regulations, permanent exemption.

Annex I: Positioning and technical details of lights and shapes

1 Definition

The term 'height above the hull' means height above the uppermost continuous deck. This height shall be measured from the position vertically beneath the location of the light.

2 Vertical positioning and spacing of lights

(a) On a power-driven vessel of 20 metres or more in length the masthead lights shall be placed as follows:
  (i) the forward masthead light, or if only masthead light is carried, then that light, at a height above the hull of not less than 6 metres, and, if the breadth of the vessel exceeds 6 metres, then at a height above the hull not less than such breadth, so however that the light need not be placed at a greater height above the hull than 12 metres;
  (ii) when two masthead lights are carried the after one shall be at least 4.5 metres vertically higher than the forward one.

(b) The vertical separation of masthead lights of power-driven vessels shall be such that in all normal conditions of trim the after light will be seen over and separate from the forward light at a distance of 1000 metres from the stem when viewed from sea level.

(c) The masthead light of a power-driven vessel of 12 metres but less than 20 metres in length shall be placed at a height above the gunwale of not less than 2.5 metres.

(d) A power-driven vessel of less than 12 metres in length may carry the uppermost light at a height of less than 2.5 metres above the gunwale. When however a masthead light is carried in addition to sidelights and a sternlight, then such masthead light shall be carried at least 1 metre higher than the sidelights.

(e) One of the two or three masthead lights prescribed for a power-driven vessel when engaged in towing or pushing another vessel shall be placed in the same position as either the forward masthead light or the after masthead light; provided that, if carried on the aftermast, the lowest after masthead light shall be at least 4.5 metres vertically higher than the forward masthead light.

(f) (i) The masthead light or lights prescribed in Rule 23(a) shall be so placed as to be above and clear of all other lights and obstructions except as described in sub-paragraph (ii).
  (ii) When it is impracticable to carry the all-round lights prescribed by Rule 27(b)(i) or Rule 28 below the masthead lights, they may be carried above the after masthead light(s) or vertically in between the forward masthead light(s) and after masthead light(s), provided that in the latter case the requirement of Section 3(c) of this Annex shall be complied with.

(g) The sidelights of a power-driven vessel shall be placed at a height above the hull not greater than three-quarters of that of the forward masthead light. They shall not be so low as to be interfered with by deck lights.

(h) The sidelights, if in a combined lantern and carried on a power-driven vessel of less than 20 metres in length, shall be placed not less than 1 metre below the masthead light.
(i) When the Rules prescribe two or three lights to be carried in a vertical line, they shall be spaced as follows:
    (i) on a vessel of 20 metres in length or more such lights shall be spaced not less than 2 metres apart, and the lowest of these lights shall, except where a towing light is required, be placed at a height of not less than 4 metres above the hull.
    (ii) on a vessel of less than 20 metres in length such lights shall be spaced not less than 1 metre apart and the lowest of these lights shall, except where a towing light is required, be placed at a height of not less than 2 metres above the hull.
    (iii) when three lights are carried they shall be equally spaced.
(j) The lower of the two all-round lights prescribed for a vessel when engaged in fishing shall be at a height above the sidelights not less than twice the distance between the two vertical lights.
(k) The forward anchor light prescribed in Rule 30(a)(i), when two are carried, shall not be less than 4.5 metres above the after one. On a vessel of 50 metres or more in length this forward anchor light shall be placed at a height of not less than 6 metres above the hull.

3 Horizontal positioning and spacing of lights

(a) When two masthead lights are prescribed for a power-driven vessel, the horizontal distance between them shall not be less than one-half of the length of the vessel but need not be more than 100 metres. The forward light shall be placed not more than one-quarter of the length of the vessel from the stem.
(b) On a power-driven vessel of 20 metres or more in length the sidelights shall not be placed in front of the forward masthead lights. They shall be placed at or near the side of the vessel.
(c) When the lights prescribed in Rule 27(b)(i) or Rule 28 are placed vertically between the forward masthead light(s) and the after masthead light(s) these all-round lights shall be placed at a horizontal distance of not less than 2 metres from the fore and aft centreline of the vessel in the athwartship direction.

4 Details of location of direction-indicating lights for fishing vessels, dredgers and vessels engaged in underwater operations

(a) The light indicating the direction of the outlying gear from a vessel engaged in fishing as prescribed in Rule 26(c)(ii) shall be placed at a horizontal distance of not less than 2 metres and not more than 6 metres away from the two all-round red and white lights. This light shall be placed not higher than the all-round white light prescribed in Rule 26(c)(i) and not lower than the sidelights.

(b) The lights and shapes on a vessel engaged in dredging or underwater operations to indicate the obstructed side and/or the side on which it is safe to pass, as prescribed in Rule 27(*d*)(i) and (ii), shall be placed at the maximum practical horizontal distance, but in no case less than 2 metres, from the lights or shapes prescribed in Rule 27(*b*)(i) and (ii). In no case shall the upper of these lights or shapes be at a greater height than the lower of the three lights or shapes prescribed in Rule 27(*b*)(i) and (ii).

5 *Screens for sidelights*

The sidelights of vessels of 20 metres or more in length shall be fitted with inboard screens painted matt black, and meeting the requirements of Section 9 of this Annex. On vessels of less than 20 metres in length the sidelights, if necessary to meet the requirements of Section 9 of this Annex, shall be fitted with inboard matt black screens. With a combined lantern, using a single vertical filament and a very narrow division between the green and red sections, external screens need not be fitted.

6 *Shapes*

(a) Shapes shall be black and of the following sizes:
   (i)   a ball shall have a diameter of not less than 0.6 metre;
   (ii) a cone shall have a base diameter of not less than 0.6 metre and a height equal to its diameter;
   (iii) a cylinder shall have a diameter of at least 0.6 metre and a height of twice its diameter;
   (iv) a diamand shape shall consist of two cones as defined in (ii) above having a common base.

(b) The vertical distance between shapes shall be at least 1.5 metre.

(c) In a vessel of less than 20 metres in length shapes of lesser dimensions but commensurate with the size of the vessel may be used and the distance apart may be correspondingly reduced.

7 *Colour specification of lights*

The chromaticity of all navigation lights shall conform to the following standards, which lie within the boundaries of the area of the diagram specified for each colour by the International Commission on Illumination (CIE).

The boundaries of the area for each colour are given by indicating the corner co-ordinates, which are as follows:

(i) *White*

x	0.525	0.525	0.452	0.310	0.310	0.443
y	0.382	0.440	0.440	0.348	0.283	0.382

(ii) *Green*

x	0.028	0.009	0.300	0.203
y	0.385	0.723	0.511	0.356

(iii) *Red*

x	0.680	0.660	0.735	0.721
y	0.320	0.320	0.265	0.259

(iv) Yellow

x	0.612	0.618	0.575	0.575
y	0.382	0.382	0.425	0.406

8 Intensity of lights

(a) The minimum luminous intensity of lights shall be calculated by using the formula:

$$I = 3.43 \times 10^6 \times T \times D^2 \times K^{-D}$$

where I is luminous intensity in candelas under service conditions,
T is threshold factor 2×10^{-7} lux,
D is range of visibility (luminous range) of the light in nautical miles,
K is atmospheric transmissivity.
For prescribed lights the value of K shall be 0.8, corresponding to a meteorological visibility of approximately 13 nautical miles.

(b) A selection of figures derived from the formula is given in the following table:

Range of visibility (luminous range) of light in nautical miles	Luminous intensity of light in candelas for $K = 0.8$
D	I
1	0.9
2	4.3
3	12
4	27
5	52
6	94

Note: The maximum luminous intensity of navigation lights should be limited to avoid undue glare.
This shall not be achieved by a variable control of the luminous intensity.

9 Horizontal sectors

(a) (i) In the forward direction, sidelights as fitted on the vessel shall show the minimum required intensities. The intensities shall decrease to reach practical cut-off between 1 degree and 3 degrees outside the prescribed sectors.

(ii) For sternlights and masthead lights and at 22.5 degrees abaft the beam for sidelights, the minimum required intensities shall be maintained over the arc of the horizon up to 5 degrees within the limits of the sectors prescribed in Rule 21. From 5 degrees within the prescribed sectors the intensity may decrease by 50 per cent up to the prescribed limits; it shall decrease steadily to reach practical cut-off at not more than 5 degrees outside the prescribed sectors.

(b) All-round lights shall be so located as not to be obscured by masts, topmasts or structures within angular sectors of more than 6 degrees, except anchor lights prescribed in Rule 30, which need not be placed at an impracticable height above the hull.

10 Vertical sectors

(a) The vertical sectors of electric lights as fitted, with the exception of lights on sailing vessels shall ensure that:
    (i)   at least the required minimum intensity is maintained at all angles from 5 degrees above to 5 degrees below the horizontal;
    (ii) at least 60 per cent of the required minimum intensity is maintained from 7.5 degrees above to 7.5 degrees below the horizontal.

(b) In the case of sailing vessels the vertical sectors of electric lights as fitted shall ensure that:
    (i)   at least the required minimum intensity is maintained at all angles from 5 degrees above to 5 degrees below the horizontal;
    (ii) at least 50 per cent of the required minimum intensity is maintained from 25 degrees above to 25 degrees below the horizontal.

(c) In the case of lights other than electric these specifications shall be met as closely as possible.

11 Intensity of non-electric lights

Non-electric lights shall so far as practicable comply with the minimum intensities, as specified in the Table given in Section 8 of this Annex.

12 Manoeuvring light

Notwithstanding the provisions of paragraph 2(*f*) of this Annex the manoeuvring light described in Rule 34(*b*) shall be placed in the same fore and aft vertical plane as the masthead light or lights and, where practicable, at a minimum height of 2 metres vertically above the forward masthead light, provided that it shall be carried not less than 2 metres vertically above or below the after masthead light. On a vessel where only one masthead light is carried the manoeuvring light, if fitted, shall be carried where it can best be seen, not less than 2 metres vertically apart from the masthead light.

13 Approval

The construction of lights and shapes and the installation of lights on board the vessel shall be to the satisfaction of the appropriate authority of the State whose flag the vessel is entitled to fly.

Annex II: Additional signals for fishing vessels fishing in close proximity

1 General

The lights mentioned herein shall, if exhibited in pursuance of Rule 26(*d*), be placed where they can best be seen. They shall be at least 0.9 metre apart but at a lower level than lights prescribed in Rule 26(*b*)(i) and (*c*)(i). The lights shall be visible all round the horizon at a distance of at least 1 mile but at a lesser distance than the lights prescribed by these Rules for fishing vessels.

2 *Signals for trawlers*

(**a**) Vessels when engaged in trawling, whether using demersal or pelagic gear, may exhibit:
  (i) when shooting their nets:
       two white lights in a vertical line;
  (ii) when hauling their nets:
       one white light over one red light in a vertical line.
  (iii) when the net has come fast upon an obstruction:
       two red lights in a vertical line.
(**b**) Each vessel engaged in pair trawling may exhibit:
  (i) by night, a searchlight directed forward and in the direction of the other vessel of the pair;
  (ii) when shooting or hauling their nets or when their nets have come fast upon an obstruction, the lights prescribed in 2(*a*) above.

3 *Signals for purse seiners*

Vessels engaged in fishing with purse seine gear may exhibit two yellow lights in a vertical line. These lights shall flash alternately every second and with equal light and occultation duration. These lights may be exhibited only when the vessel is hampered by its fishing gear.

Annex III: Technical details of sound signal appliances

1 *Whistles*

(**a**) *Frequencies and range of audibility*

The fundamental frequency of the signal shall lie within the range 70–700 Hz.

The range of audibility of the signal from a whistle shall be determined by those frequencies, which may include the fundamental and/or one or more higher frequencies, which lie within the range 180–700 Hz ($\pm$ 1 per cent) and which provide the sound pressure levels specified in paragraph 1(*c*) below.

(b) *Limits of fundamental frequencies*
To ensure a wide variety of whistle characteristics, the fundamental frequency of a whistle shall be between the following limits:
    (i)   70–200 Hz, for a vessel 200 metres or more in length;
    (ii) 130–350 Hz, for a vessel 75 metres but less than 200 metres in length;
    (iii) 250–700 Hz, for a vessel less than 75 metres in length.

(c) *Sound signal intensity and range audibility*
A whistle fitted in a vessel shall provide, in the direction of maximum intensity of the whistle and at a distance of 1 metre from it, a sound pressure level in at least one 1/3rd-octave band within the range of frequencies 180–700 Hz ($\pm$ 1 per cent) of not less than the appropriate figure given in the table below.

Length of vessel in metres	1/3rd-octave band level at 1 metre in dB referred to 2×10^{-5} N/m²	Audibility in range in nautical miles
200 or more	143	2
75 but less than 200	138	1.5
20 but less than 75	130	1
Less than 20	120	0.5

    The range of audibility in the table above is for information and is approximately the range at which a whistle may be heard on its forward axis with 90 per cent probability in conditions of still air on board a vessel having average background noise level at the listening posts (taken to be 68 dB in the octave band centred on 250 Hz and 63 dB in the octave band centred on 500 Hz).

    In practice the range at which a whistle may be heard is extremely variable and depends critically on weather conditions; the values given can be regarded as typical but under conditions of strong wind or high ambient noise level at the listening post the range may be much reduced.

(d) *Directional properties*
The sound pressure level of a directional whistle shall be not more than 4 dB below the prescribed sound pressure level on the axis at any direction in the horizontal plane within $\pm$ 45 degrees of the axis. The sound pressure level at any other direction in the horizontal plane shall be not more than 10 dB below the prescribed sound pressure level on the axis, so that the range in any direction will be at least half the range on the forward axis. The sound pressure level shall be measured in that 1/3rd-octave band which determines the audibility range.

(e) *Positioning of whistles*
When a directional whistle is to be used as the only whistle on a vessel, it shall be installed with its maximum intensity directed straight ahead.

    A whistle shall be placed as high as practicable on a vessel, in order to reduce interception of the emitted sound by obstructions and also to minimize hearing damage risk to personnel. The sound pressure level of the vessel's own signal at listening posts shall not exceed 110 dB (A) and so far as practicable should not exceed 100 dB (A).

(f) *Fitting of more than one whistle*
If whistles are fitted at a distance apart of more than 100 metres, it shall be so arranged that they are not sounded simultaneously.

(g) *Combined whistle systems*
If due to the pressure of obstructions on the sound field of a single whistle or of one of the whistles referred to in paragraph 1(*f*) above is likely to have a zone of greatly reduced signal level, it is recommended that a combined whistle system be fitted so as to overcome this reduction. For the purposes of the Rules a combined whistle system is to be regarded as a single whistle. The whistles of a combined system shall be located at a distance apart of not more than 100 metres and arranged to be sounded simultaneously. The frequency of any one whistle shall differ from those of the others by at least 10 Hz.

2   *Bell or gong*

(a) *Intensity of signal*
A bell or gong, or other device having similar sound characteristics shall produce a sound pressure level of not less than 110 dB at a distance of 1 metre from it.

(b) *Construction*
Bells and gongs shall be made of corrosion-resistant material and designed to give a clear tone. The diameter of the mouth of the bell shall be not less than 300 mm for vessels of 20 metres or more in length, and shall be not less than 200 mm for vessels of 12 metres or more but of less than 20 metres in length. Where practicable, a power-driven bell striker is recommended to ensure constant force but manual operation shall be possible. The mass of the striker shall be not less than 3 per cent of the mass of the bell.

3   *Approval*

The construction of sound signal appliances, their performance and their installation on board the vessel shall be to the satisfaction of the appropriate authority of the State whose flag the vessel is entitled to fly.

Annex IV: Distress Signals (see page 159)

Appendix N
Port Traffic Signals

Proposed Unification of Port Traffic Signals 1983

	Lights	Meaning
1.	Ⓡ Ⓡ Ⓡ	Serious emergency – all vessels to stop or to divert according to instructions. *In this case the lights are flashing*
2.	Ⓡ Ⓡ Ⓡ	Vessels shall not proceed
2a.	Ⓨ Ⓡ Ⓡ Ⓡ	Vessels shall not proceed, except that vessels which navigate outside the main channel need not comply with the main message
3.	Ⓖ Ⓖ Ⓖ	Vessels may proceed. One way traffic
4.	Ⓖ Ⓖ Ⓦ	Vessels may proceed. Two way traffic
5.	Ⓖ Ⓦ Ⓖ	A vessel may proceed only when she has received specific orders to do so
5a.	Ⓨ Ⓖ Ⓦ Ⓖ	A vessel may proceed when she has received specific orders to do so, except that vessels which navigate outside the main channel need not comply with the main message.

Additional signals may be added, if necessary, for special purposes but normally to the right of the column carrying the main message. No additional lights should be added to the main message. Red indicates 'Do not proceed'. Green indicates 'Proceed, subject to conditions.' The single yellow light to the left of the column signifies that vessels which can navigate outside the main channel need not comply with the main message. Except at 1, lights shown are fixed or slow occulting.

DISTRESS AND URGENCY by VHF radio

DISTRESS A distress message indicates the yacht is threatened by grave and imminent danger and requests immediate assistance. It has priority over all other transmissions.

'MAYDAY   MAYDAY   MAYDAY'
'THIS IS – Yacht's name   Yacht's name   Yacht's name'

Followed by the message which begins: Mayday – Yacht's name
Followed by the vessel's position, the nature of the distress and any other important information to facilitate rescue.
Followed by:

'OVER' (the invitation to reply)

Then release the 'Press-to-speak' switch and listen.

URGENCY A very urgent message concerning the safety of a ship or person has priority over all other communications, except distress, and should be preceded by:

'PAN PAN   PAN PAN   PAN PAN'
'HELLO ALL STATIONS   ALL STATIONS   ALL STATIONS'
'THIS IS – Yacht's name   Yacht's name   Yacht's name'

Followed by the message which should begin with the yacht's position (given as for a distress message) and the assistance required.
Followed by:

'OVER'   Then listen.

ASSISTANCE TO OTHERS If you see or hear a distress signal you must proceed with all speed to the assistance of the persons in distress. Switch to Channel 16 and listen. If your transmission will not interfere with distress traffic, call the Coastguard, or the nearest Coast Radio Station, to report your position, what you have seen and your intentions. Then remain tuned to Channel 16.
If it is necessary to speak to other vessels in the vicinity and after calling and receiving an acknowledgement on Channel 16, transfer to Channel 6 for scene-of-search intership traffic.

SAFETY In the UK Channel 67 is used for messages between HM Coastguard and small craft on non-immediate ship safety or navigational safety matters. Call the Coastguard on Channel 16 and say you have a safety message. You will be asked to transfer to Channel 67 to pass your message.